DECEPTION IN WAR

DECEPTION IN WAR

JON LATIMER

THISTLE
PUBLISHING

Copyright © Jon Latimer 2001, 2015

First published 2001 by John Murray

This edition published in 2015 by:

Thistle Publishing
36 Great Smith Street
London
SW1P 3BU

www.thistlepublishing.co.uk

ISBN-13: 978-1-910670-67-5

To my father, who started it all . . . and to C Coy, 3 (V) RWF, and all who served in it

CONTENTS

ILLUSTRATIONS

The author and publishers would like to thank the following for permission to reproduce illustrations: Plate 1, Carmarthenshire County Museum; 2, Library of Congress; 3, 4, 8, 9, 10, 11, 14, 15, 16, 17, 19, 20, 21, 22, 23, 24 and 25, Imperial War Museum; 5, John Murray Archive; 6, 7 and 12, Tank Museum; 13 and 18, Public Record Office; 26, British Army; 28, TVI Corporation; 29, Saab Barracuda AB.

PREFACE

lthough this work is intended for the general reader
rather than as an academic tome or military manual,
I make no apology for the inclusion of endnotes. In
his excellent *The Atlantic Campaign* Dan van der Vat ex-
presses a violent dislike for texts 'bespattered by numbers',
but these notes are often of considerable value to anyone
wishing to pursue specific points. They are all related to
the source material, and the few additional footnotes included merely provide details that do not properly belong
in the main text.

A great many people have contributed to this book
and I would also like to thank others who have contributed elsewhere in the meantime. In particular, my old
friend Marcus Bennett, police officer and captain in the
Royal Welsh Regiment, has been a veritable mother lode
of ideas and help; similarly John Hall, lecturer in Spanish
at UW Swansea and former intelligence officer with 4
RRW, has provided me with copious leads on source
material and excellent advice, as have Ally Morrison,
Major James Everard QRL, Martin Coulson, (also a lecturer at UW Swansea and former commanding officer,

R Mon RE (M)), Nick Pope, Marcus Cowper, George Forty and David Nicolle.

The staff at the city library, Swansea, and UW Swansea have patiently filed all my inter-library loan requests; the staff at the Imperial War Museum and David Fletcher and the staff at the Tank Museum library, and Jillian Brankin at the Australian War Memorial have been ever friendly and helpful over the past year. Jon Guttman at *Military History* magazine has been very patient and helpful, and thanks are also due to the staff of the other titles at Primedia History Group. Thanks also to Lee Johnson at Osprey and especially to John McHugh, Kevin Enright and Christopher Samuel for putting up with me when in London. I want to thank my agent, Andrew Lownie, for making it all happen (and Adrian Weale who – albeit inadvertently – put us in touch), Grant McIntyre and everyone at John Murray, and Matthew Taylor, who edited the typescript meticulously.

Swansea
November 2000

MAPS

NORTH SEA

UNITED
NETHERLANDS
• Rotterdam

OVERKIRK

XXXX

• Antwerp
MARLBOROUGH

XXXX

SPANISH
NETHERLANDS   • Cologne
• Brussels

Ramillies   Meuse   • Liège   Bedburg

Namur   SAXONY

VILLEROI   XXXX

• Sedan

Koblenz

Moselle   • Frankfurt

• Mainz   Main

PALATINATE

Thionville   • Mannheim • Heidelburg

• Metz   • Philippsburg

LORRAINE   XXXX

TALLARD

Rhine   Neckar   Nördlingen • Donauwörth

BADEN   Ingolstadt
• Strasbourg   BATTLE OF BLENHEIM
13 AUGUST 1704

WÜRTTEMBERG   • Ulm

Vosges   Danube   XXXX   BAVARIA

ALSACE   ELECTOR   • Munich
OF BAVARIA

FRANCE   Black Forest   • Vienna

• Belfort   AUSTRIA

Jura   TYROL

0   miles   100
0   kilometres   160

MARLBOROUGH'S GRAND DECEPTION:
THE MARCH TO THE DANUBE, 1704

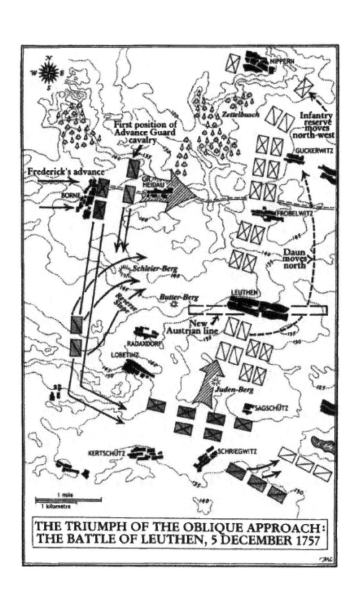

THE TRIUMPH OF THE OBLIQUE APPROACH:
THE BATTLE OF LEUTHEN, 5 DECEMBER 1757

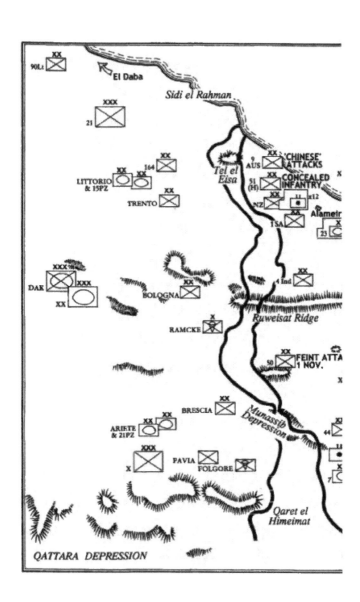

90L₁ XX

El Daba

Sidi el Rahman

21 XXX

164 XX

LITTORIO & 15PZ XX

TRENTO XX

Tel el Eisa

9 AUS XX 'CHINESE' ATTACKS

51 (H) XX CONCEALED INFANTRY

NZ XX 11 ⊕ x12

Alamein

TSA XX

23 X C

DAK XXX 'N

XXX

XX

BOLOGNA XX

4 Ind XX

Ruweisat Ridge

RAMCKE X

50 XX FEINT ATTA 1 NOV.

X

BRESCIA XX

Munassib Depression

ARIETE & 21PZ XX

X XXX

PAVIA XX

FOLGORE X

44 X

18 ⊕

7 X

Qaret el Himeimat

QATTARA DEPRESSION

OPERATION BERTRAM : EL ALAMEIN, OCTOBER 1942

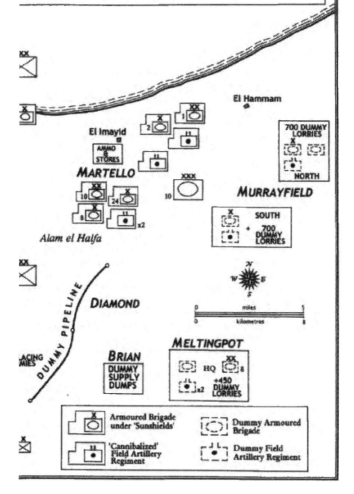

El Hammam

El Imayid

AMMO & STORES

MARTELLO

Alam el Halfa

700 DUMMY LORRIES

NORTH

MURRAYFIELD

SOUTH

+ 700 DUMMY LORRIES

DIAMOND

DUMMY PIPELINE

ACING MIES

BRIAN

DUMMY SUPPLY DUMPS

MELTINGPOT

HQ

+450 DUMMY LORRIES

N
W   E
S

miles

kilometres

Armoured Brigade under 'Sunshields'

'Cannibalized' Field Artillery Regiment

Dummy Armoured Brigade

Dummy Field Artillery Regiment

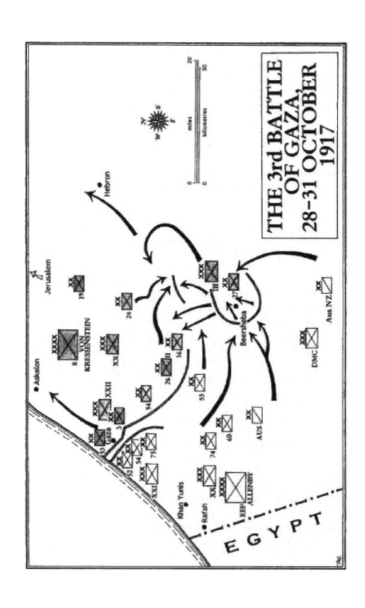

THE 3rd BATTLE
OF GAZA,
28–31 OCTOBER
1917

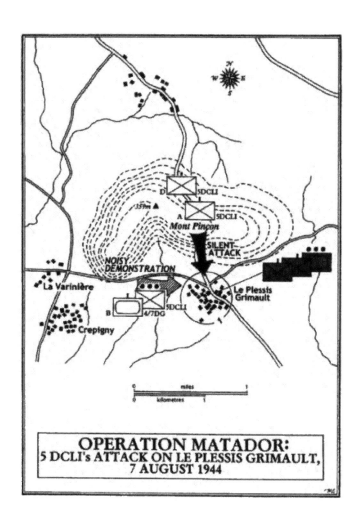

OPERATION MATADOR:
5 DCLI's ATTACK ON LE PLESSIS GRIMAULT,
7 AUGUST 1944

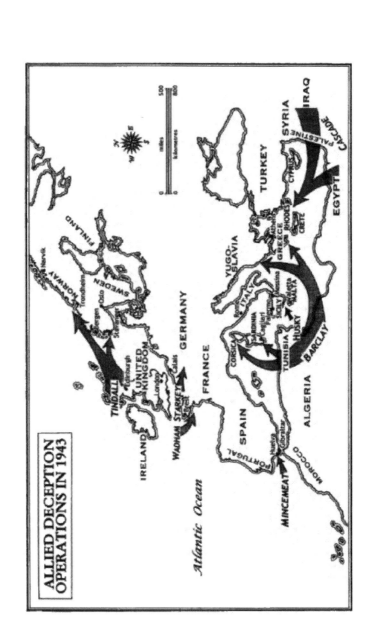

ALLIED DECEPTION
OPERATIONS IN 1943

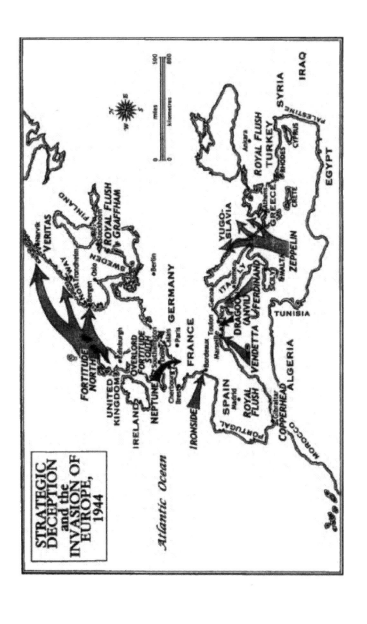

STRATEGIC
DECEPTION
and the
INVASION OF
EUROPE,
1944

Atlantic Ocean

FORTITUDE NORTH
VERITAS
VaÂ
Narvik
NORWAY
Oslo
SWEDEN
FINLAND
ROYAL FLUSH
GRAFFHAM
Stockholm
Trondheim

UNITED KINGDOM
IRELAND
Edinburgh
OVERLORD
FORTITUDE SOUTH
NEPTUNE
Cherbourg
Brest
Caen
Paris
FRANCE
GERMANY
Berlin

IRONSIDE
Bordeaux
Toulon
Marseille

SPAIN
Madrid
ROYAL FLUSH
PORTUGAL
Gibraltar
MOROCCO
COPPERHEAD
ALGERIA
TUNISIA

ITALY
Rome
DRAGOON
(ANVIL)
VENDETTA
FERDINAND

YUGO-SLAVIA
ZEPPELIN
SPALTO
Split

GREECE
ROYAL FLUSH
Argos
RHODES
CRETE
TURKEY
SYRIA
CYPRUS
PALESTINE
IRAQ
EGYPT

miles
kilometres
500
800

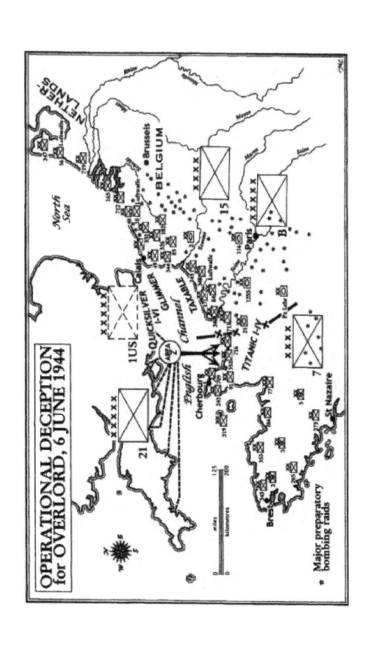

OPERATIONAL DECEPTION
for OVERLORD, 6 JUNE 1944

NETHER-LANDS

North Sea

Brussels

BELGIUM

15

B

Paris

Calais

QUICKSILVER I–VI

GLIMMER

TAXABLE

English Channel

1USL

AREA Z

Cherbourg

TITANIC I–IV

21

7

Brest

St Nazaire

* Major preparatory
  bombing raids

miles
kilometres

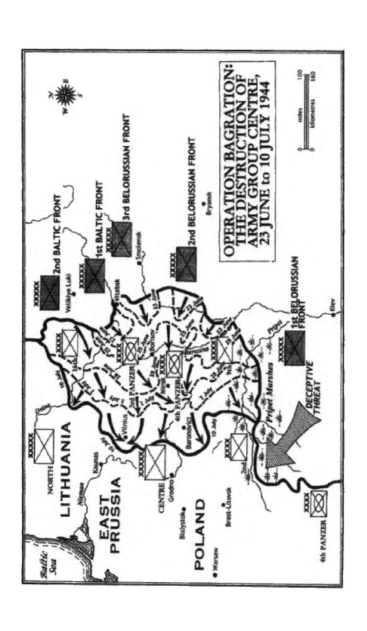

OPERATION BAGRATION:
THE DESTRUCTION OF
ARMY GROUP CENTRE,
23 JUNE to 10 JULY 1944

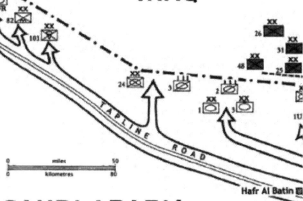

As Samawah

OPERATION DESERT STORM: THE 'HAIL MARY' PLAY, 24 FEBRUARY 1991

As Salman

Al Busayya

ordinary Iraqi units

Republican Guard

IRAQ

TAPLINE ROAD

Hafr Al Batin

SAUDI ARABIA

miles

kilometres

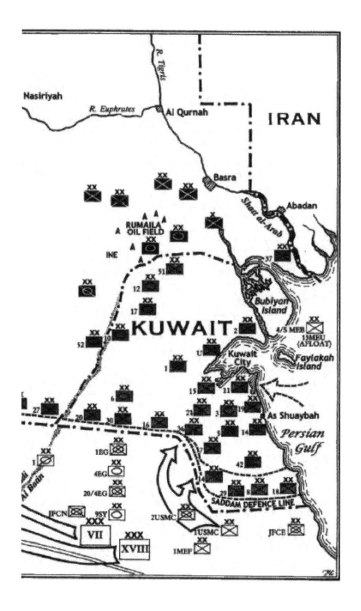

Messieurs les maréchaux Murat, Lannes, and Belliard get on their horses and ride down to the bridge. (Observe that all three are Gascons.) 'Gentlemen', says one of them, 'you are aware that the Thabor bridge is mined and doubly mined, and that there are menacing fortifications at its head and an army of fifteen thousand men has been ordered to blow up the bridge and not let us cross? But it will please our Sovereign the Emperor Napoleon if we take this bridge. So let us three go and take it!' 'Yes, let's!' say the others ...

These gentlemen ride on to the bridge alone, and wave white handkerchiefs; they assure the officer on duty that they, the marshals, are on their way to negotiate with Prince Auersperg. He lets them enter the *tête du pont* [bridgehead]. They spin him a thousand gasconades, saying that the war is over, that the Emperor Francis is arranging a meeting with Bonaparte, that they desire to see Prince Auersperg, and so on. The officer sends for Auersperg; these gentlemen embrace the officers, crack jokes, sit on the cannon, and meanwhile a French battalion gets to the bridge unobserved, flings the bags of incendiary material into the water, and approaches the *tête du pont*. At length appears the lieutenant-general, our dear Prince Auersperg von Mautern himself. 'Dearest foe! Flower of the Austrian army, hero of the Turkish wars! Hostilities are ended, we can shake one another's hand ... The Emperor Napoleon burns with impatience to make Prince Auersperg's acquaintance.' In a word those gentlemen, Gascons indeed, so bewilder him with fine words, and he is so flattered by his rapidly established intimacy with the French marshals, and so dazzled by the sight of Murat's mantle and ostrich plumes, that their fire gets into his eyes and he forgets that he ought to be firing at the enemy. The French battalion rushes to the bridgehead, spikes the guns and the bridge is taken! But what is best of all is that the sergeant in charge of the cannon which was to give the signal to fire the mines and blow up the bridge, this sergeant seeing that the French troops were running onto the bridge, was about to fire, but Lannes stayed his hand. The sergeant, who was evidently wiser than his general, goes up to Auersperg and says: 'Prince, you are being deceived, here are the French!' Murat, seeing that all is lost if the sergeant is allowed to speak, turns to Auersperg with feigned astonishment (he is a true Gascon) and says: 'I don't recognize the world-famous Austrian discipline, if you allow a subordinate to address you like that!' It was a stroke of genius. Prince Auersperg feels his dignity at stake and orders the sergeant to be arrested. Come, you must own that this affair of the Thabor bridge is delightful!

Leo Tolstoy, *War and Peace*

'All warfare is based on deception.
Therefore, when capable, feign incapacity;
when active, inactivity ...
Offer the enemy a bait to lure him;
feign disorder and strike him ...
Pretend inferiority and encourage his arrogance.'

Sun Tzu

INTRODUCTION

'Partout la violence produit la ruse.'
Bernardin de Saint-Pierre

'SURPRISE IS A Principle of War ... [It] should primarily be directed at the mind of an enemy commander rather than at his force. The aim should be to paralyse the commander's will.'[1] Surprise is the great 'force multiplier' – it makes one stronger than is physically the case. Surprise can be achieved by a variety of methods: by forgoing preparations that an enemy might expect one to make, by attacking at an unexpected time, by using ground deemed impassable (as with the German drive through the Ardennes in May 1940), through bold and innovative tactics or by the employment of powerful new weapons (the T-34 tank came as a terrible surprise to the Germans in the USSR in 1941). However, among the many factors contributing to the achievement of surprise, surely the most important is deception.[2]

It might be argued that security is an even more important concern, but in battle it is not sufficient for a commander to avoid error; he needs actively to cause his enemy to make mistakes.[3] Deception is an active measure with precisely that aim (requiring security among other things and including passive elements such as camouflage), and since the stratagem, or *ruse de guerre,* is as old as warfare itself, it is a foolish commander who ignores it. Indeed, the greatest generals in history have been masters of it, and it has been the downfall of many another.

Everybody employs deception at times, either to gain an advantage or for more altruistic reasons. Although adults reprimand their children for lying, they themselves lie all the time, especially to their children. Deception is such an integral part of our lives that we often fail to recognize it. Surveys indicate that politicians are distrusted because they are perceived as deceitful, but everyone recognizes that a certain measure of 'economy with the *actualité*' is a necessary requirement of the profession. If politicians always said exactly what they thought, they would have very short careers.

There is, as the saying goes, nothing new under the sun, and as we examine the historical development of deception in war we will see the same themes and techniques recurring and repeating themselves in subtle new ways. However, this book is intended not as a history of deception – that would be a lifetime's work – but as an examination of the *art* of deception. To be successful, the deceiver needs to know and understand the mind of the enemy commander.

> Rashness, excessive audacity, blind impetuosity or foolish ambition are all easily exploited by

the enemy and most dangerous to any allies, for a general with such defects in his character will naturally fall victim to all kinds of stratagems, ambushes and trickery.[4]

The place of self-deception in this process is an important one. Our perceptions develop through the process of learning, but are overlain by a sociological and cultural baggage that correlates to our prejudices. Much of the time we view our experiences through these mental templates, and whatever does not fit our prejudices tends to be overlooked or discarded.

The elders in most societies have traditionally been regarded as the repositories of collective wisdom, which tends to reinforce conservatism in thought – a particular tendency in the military that Norman Dixon highlights in his book *On the Psychology of Military Incompetence.* Under stress this tendency tends to be further reinforced. We hate disorder and confusion and our every mental effort tries to impart order and meaning to events; even when information is limited or contradictory, we remain eager to draw conclusions. And since the mind can only cope with so much information at any one time, we are forced to filter and prioritize the information stream. Thus it can be said that all deception in war should be based on what the enemy himself not only believes, but hopes for.[5]

Information is a premium commodity on any battlefield and increasingly vast amounts of it are required for successful operations, with many means being employed to collect and process it. Consequently, skilfully conveyed false information often has great influence on the mind of an enemy and the course of operations. Since military organizations look through doctrinal and physical templates as well as the mental templates of its individual members, it is this that provides the basis for deception. The information an enemy requires to make decisions can be manipulated, if one understands the templates he is using. And a reputation for being crafty and deceptive will enhance the anxiety and uncertainty of one's opponent.[6]

War is the most extreme condition that most people are ever likely to face. It is not a 'gentlemanly' pursuit but often a matter of survival requiring ruthless measures in its pursuit. So it is very often in times of weakness that commanders first think of deception as a means of evening the odds. The Marxist-Leninist system, with its belief in inevitable and predictable dialectical change, accepted that anything that promoted that change was desirable if not essential, and that deception was therefore a legitimate tool in peace and war (as the Soviets demonstrated between 1941 and 1945). In the West, on the other hand, deception is often seen as immoral, and more than one authority has claimed that, as a result, Americans resort to deception only reluctantly or else do it poorly.[7] In fact, however, many Americans displayed a natural flair for deception during the Civil War, just as they had during the Revolution eighty years previously.

Yet for a long time deception did indeed run counter to the American concept of military honour. There was a strange reluctance among some Americans during the twentieth century to accept it as part of modern warfare, and certainly the Americans resorted to deception intermittently during the Second World War.[8] Colonel William A. Harris, the principal American deception officer in Europe, was converted to belief in the value of FORTITUDE SOUTH (part of the deception cover plan for the Normandy landings in 1944) only after its success.[9] Perhaps by this stage the Americans felt sufficiently strong to win the war without resort to deception, whereas earlier on, when they were weak as a result of the Japanese attack on Pearl Harbor, the US Navy made extensive use of it. Yet deception can clearly be benign: were it not for BODYGUARD (the overall plan of which FORTITUDE was a part) the defeat of the Third Reich would undoubtedly have been yet more protracted and bloody.

The British, despite their reputation for 'fair play', have long shown a remarkable flair for deception, and by the end of the Second World War they had an unrivalled mastery in the art of military deception (which they have since largely forgotten). In contrast, the most efficient military machine of the past century – that of Germany – has been less strongly inclined towards deception, except in the form of Hitlerian machinations. While the German Army has always understood the importance of surprise and has consistently achieved it, its preferred method has generally been the one described by Frederick the Great as 'speed and violence'.

The purpose of this book is to describe and explain the systematic telling of lies for specifically military purposes. In this context we are dealing with very creative minds that seek to weave delicate tapestries of information in a fragile and hostile environment. It is a difficult process that combines great risk with the potential for enormous gain. The most effective deceivers display an unorthodoxy of thought that is usually little appreciated in a peacetime army. Perhaps more than any other branch of military endeavour, successful deception is an art rather than a science, although science increasingly provides the technical means by which deception is created. Many of the best practitioners have had backgrounds in both the visual and the performing arts, but the art of deception is most successful when applied patiently, with proven techniques guided by solid principles. These we will examine in the light of examples from history, but with particular reference to the twentieth century, when technology transformed the techniques, if not the principles of deception, and thus complicated matters considerably. Some would say that modern technology renders deception more difficult but throughout history deceivers have exploited the latest technological developments. The information revolution taking place today is having an impact comparable to that of the industrial revolution and will probably be accompanied by changes on a similar scale in the nature of war; but deception will no doubt continue just as long as warfare does.

And, of course, there is something deliciously, wickedly, entertaining about pulling the wool over an opponent's eyes. Welcome to a book packed with such lies.

CHAPTER ONE

A HISTORY OF BLUFF
IN WARFARE

'But now change your theme and sing to us of the
stratagem of the Wooden Horse, which Epeius
built with Athene's help, and which the good
Odysseus contrived to get taken one day into the
citadel of Troy as an ambush, manned by the warriors who then sacked the town.'

Homer

DECEPTION IN ANCIENT AND MEDIEVAL WARFARE

DECEPTION ON THE battlefield is surely as old as warfare
itself. One of the most famous early examples dates from
c. 1294 BC, when Pharaoh Ramses II of Egypt led his army
against the Hittite stronghold of Kadesh. Two Hittite
'deserters' came to him offering to lead him against their
former comrades. Instead, they led him into an ambush
that very nearly proved disastrous.

Some 400 years later and not far away, ancient Israel
was overrun by the Midianites (nomadic Arab tribesmen
who regularly brought their flocks to graze the lowlands
where the Israelites had sown their crops). Gideon, son

of Joash, resolved to drive them off. In seven previous years the Israelites had hidden in the hills on the approach of the Midianites, and it was with difficulty that Gideon assembled just 300 men for the task. Only guile could achieve what numbers could not. Gideon first took care to ensure that tales of signs and portents marking the rise of a great new Israelite leader filtered down to the Midianite camp. Then each man was issued with a trumpet, a pitcher and a torch. The torches were lit and carefully concealed under the pitchers, and, with their trumpets in their hands and divided into three companies, the 300 took up positions around the enemy camp. At around midnight, when the Midianites were known to change their sentries, Gideon's men gave out an almighty cry – 'The sword of the Lord and of Gideon!' – accompanied by loud blasts of trumpets and the waving of hundreds of torches. The Midianites, convinced that they were being attacked by a great host, were sent tumbling in panic for the fords on the River Jordan, harried all the way by the Israelite population, which rose *en masse* now its enemies were on the run. Gideon relentlessly pursued them to ensure the full exploitation of his success, and 'the day of Midian' became a proverb in Israel for total victory.[1]

The name of Sun Tzu is nowadays synonymous with the idea of deception. His *Art of War* has been a key reference source for Chinese strategists and military leaders for over 2,000 years, although it was properly translated into English only at the beginning of the twentieth century. The exploits of the Ch'i general Sun Pin in 341 BC provide an interesting example of the theories of Sun Tzu in combat. Before his invasion of the territory of Wei, Sun Pin assessed the situation with an advisor, who said: 'The soldiers of Wei are fierce and bold, and despise the men of Ch'i as cowards. A skilful strategist should make use of this and lure them with the promise of advantage ... [L]et us light a hundred thousand fires when our army enters Wei, fifty thousand the next day, and only thirty thousand on the third day ...', thereby indicating to the Wei general P'ang Chuan that the army of Ch'i was experiencing mass desertions and encouraging him to rush to the attack. P'ang Chuan took the bait and led his forces through a narrow gorge preselected by Sun Pin for the ambush. As a final finesse Sun Pin posted a sign. When he arrived at the ambush site, P'ang Chuan called for a torch to read Sun Pin's sign, which said: 'P'ang Chuan dies beneath this tree.' The lighting of the torch was the signal for Sun Pin's archers to shoot.[2]

By virtue of the serious nature of war, it may sometimes be justifiable and even necessary to deceive one's own side. During the march from Spain to Italy the great Carthaginian general Hannibal Barca, probably the greatest exponent of deception in the classical world, found it necessary to deceive his own elephants. His army had to cross the River Rhône, but the elephants accompanying it would on no account enter the water. So Hannibal's pioneers built rafts, two of which were firmly lashed together on the bank, with further rafts then added to form a pontoon projecting some 200 feet into the water and made absolutely fast against the bank. Two more rafts were then added at the end of the pier with towing lines to boats in the river, but with lashings to the pier that could easily be cut. The whole pier was then covered with earth to make it appear like an extension of the bank and two female elephants led the way – to encourage the others. When the elephants were standing on the final rafts the lashings were cut, and once they found themselves in midstream, the elephants had little option but to complete the crossing. The process was repeated, and although a few elephants tipped into the river in panic, they swam the rest of the way and the operation was successful.[3]

Of course, it is more common for opposition from the enemy to makes deception imperative. During the rebellion of Vercingertorix in Gaul in 52 BC Julius Caesar was marching into the country of the Arverni towards the town of Gergovia, following the course of the Allier, a wide river that flows into the Loire near Nevers. Vercingertorix broke up all the bridges across the Allier and marched his force along the opposite bank, keeping Caesar in view and planning to contest any attempted crossing. He placed patrols wherever the Romans might try to build a bridge, and it seemed that Caesar would be held up all summer since the river was not normally fordable until the autumn. Caesar camped in woods near one of the broken bridges for the night and the following morning instructed two legions to remain concealed there; he then broke the other four legions down into companies to give the appearance that all six legions were marching, and sent them with the entire baggage train to march as far as they could. Having waited for them to get clear, Caesar then emerged from hiding and quickly rebuilt the bridge on the original piles, which were still intact. The legions then formed a bridge-head on the far bank and Caesar recalled the main body. Shocked, Vercingertorix marched away to Gergovia.[4]

Deception was such a common aspect of ancient warfare that when Julius Sextus Frontinus wrote two volumes on the art of war in the first century AD (the first of which is now lost), the second volume, called *Stratagems*, was entirely devoted to the subject. In four books Frontinus describes all manner of military tricks and sleights of hand from the ancient world. Yet publicly the Romans showed a haughty contempt for such tactics.

During the early Middle Ages the Western creed of chivalry frowned upon deception, which, since most battles were fought at close quarters, appeared in any case to have limited application.[5] Further east, however, war and deception were studied as an art for centuries after the fall of Rome. Indeed, the Byzantines suffered not even the tiniest hint of chivalric sentiment, but had rather a burgeoning professional pride in their skill at deception. Among the greatest of all the soldiers of this period was the Byzantine general Belisarius. A superb fighter and trainer of men, he served his ungrateful master, Emperor Justinian, with unswerving loyalty and skill. The parsimonious emperor frequently entrusted Belisarius with difficult missions but never allocated him the resources to achieve them. Deception is often the last resort of commanders in positions of weakness and Belisarius was always considering ways to outwit his opponent by stratagem as much as by fighting.[6] Other Byzantine leaders also saw deception as being perfectly natural in warfare.

5

They considered it absurd to spend blood and treasure on achieving their aims if these could be achieved by skill, and thus developed a strong predilection for ruses, stratagems and feigned retreats. In his *Tactica* Emperor Leo VI demonstrates no shame in some of the over-ingenious stratagems used, and recommends one trick in particular that remained in use into the twentieth century – that of writing treasonable letters to officers in the enemy camp and ensuring they fall into the wrong hands. He also goes on to describe how nothing worked better against the Franks and Lombards than a feigned flight, which they always followed hastily.[7]

It is likely that the Normans learned from the Byzantines this tactic of the feigned retreat. Norman adventurers first settled in Sicily in 1016 and established a permanent stronghold at Aversa. The Byzantine army that invaded eastern Sicily in 1038 included many Normans, who served as mercenaries in a number of armies and who subsequently spread all over southern Italy. In 1060 Robert Guiscard (whose name meant 'wily' in Norman French) began the Norman conquest of Sicily, which included a prolonged campaign against the Byzantines. Shortly afterwards, Duke William of Normandy invaded England to seek its crown. The English under King Harold occupied a strong position along a hilltop near Hastings, and after the Norman archers failed to make an impression on the English line, the initial assaults by heavily armoured cavalry and foot soldiers were also repulsed. William of Poitiers then states that the Normans, 'realizing that they could not overcome an enemy so numerous without great loss to themselves ... retreated, deliberately feigning flight'. The Breton cavalry on the left of the Norman line were definitely the first to break, and many of the remaining troops followed suit, believing Duke William to be dead, but he quickly rode along the line and rallied it before turning on a party of English that had followed the Bretons and destroyed them. He then renewed the assault on the main English position. All the contemporary sources refer to this ruse repeatedly drawing groups of English in pursuit, whereupon they were destroyed piecemeal. Although this tactic had already been used by the Normans at Arques in 1053 and Messina in 1060, scholars have long continued to debate the veracity of these reports.[8]

Hans Delbrück insisted that a feigned flight was beyond the capabilities of medieval cavalry.[9] On the other hand, Sir Charles Oman had no doubt that 'a sudden inspiration came to William ... After all, Guy of Amiens, an absolute contemporary, describes it clearly.'[10] More reasonably, Hastings was probably too disjointed a battle for the necessary control of a feigned retreat to be exercised all along the Norman line, and it is perhaps more likely that local withdrawals drew groups of defenders from their positions in a series of retreats and counter-charges. Whatever the truth, the battle has since earned a reputation as an example of masterful tactical deception.

A feigned withdrawal would undoubtedly be a difficult manœuvre to achieve in battle, since it would put the troops involved at great risk. Nevertheless, the Saracens would often try to feign withdrawal while fighting the Crusaders, sometimes for days on end, in order to draw their more heavily armed opponents onto favourable ground. The feigned withdrawal was also a favourite tactic of the Mongols. A light cavalry corps of 'suicide troops' called the *mangudai* existed for the purpose (the name was not so much a job description as a tribute to the soldiers' bravery). They would charge the enemy alone, break ranks and run in an attempt to lure the enemy to destruction. The larger the *mangudai,* the more effective would be the lure: where the ground was open and favourable, it could comprise up to half the army. If the enemy did give chase, they would find themselves showered with arrows; once the quivers were emptied, the heavy cavalry would charge, always the final stage in the Mongol battle plan, delivered at the trot and in silence until the order to gallop was given at the last possible moment. As the Muscovites found to their cost at the Kalka River in 1223, the result was absolutely devastating.[11]

The Mongols would gladly use any means to gain an advantage, and many of their inspirational expedients were produced by allowing junior commanders to use their initiative. As soon as the plan of campaign had been agreed at the *kuriltai* (the great council of war), rumours would be deliberately planted exaggerating the numbers of their army. This simple and effective deception was then given credibility by the Mongols' extreme manœuvrability and speed, as demonstrated in their campaigns against the Khwarezms in Central Asia, in which an army of more than 200,000 men, operating in four corps across a 200-mile frontage, introduced a scale and speed of warfare not seen again until Napoleon's day.

The Mongols could strike terror into their opponents by appearing in strength in different places at the same time, and since each Mongol went on campaign with a number of horses (the numbers quoted vary, but five per man seems reasonable), the mounting of dummy riders on spare horses enabled them to multiply their apparent numbers further.[12]

The Mongols liked to operate during the winter, when they would be able to cross frozen marshes and rivers. To find out if the ice would support them, they would encourage the local population to test it. In Hungary in late 1241 the Mongols left cattle unattended on the left bank of the Danube in sight of starving refugees they had driven across the river earlier in the year. When the Hungarians crossed the river to recover the cattle, the Mongols swiftly followed up. Another common Mongol ploy was the use of smokescreens (used by the Greeks as early as the Pelopponesian Wars, c. 431-404 BC), by sending out small detachments to light enormous prairie fires or shooting containers of burning tar from their improvised artillery. At the Battle of Liegnitz in 1241 they set fire to reeds, and on other occasions they would light fires in inhabited regions in order to deceive the enemy as to their real intentions and to cover their movements.[13]

By the middle of the thirteenth century the Crusader states of the Middle East found themselves squeezed between the Mongol conquerors of Persia and the Mameluke Empire of Egypt. As the Mongol tide receded from Syria, so the Mameluke Sultan Baybars finally captured the great Crusader fortress of Crac des Chevaliers from the Knights Hospitaller in 1271. Before the use of gunpowder became widespread, a castle of such power could be taken only by starvation or trickery. Baybars commenced his siege between 18 and 21 February and managed to storm the forward defences and the barbicans. But the main keep or *donjon* was practically impregnable, and Baybars realized it could be taken only with heavy losses or a prolonged siege. Instead, he passed a forged letter into the keep in which the Knights' commander ordered the garrison to surrender. Whether they fell for the trick or were merely aware of the helplessness of their position, the Knights complied, despite having successfully resisted all previous sieges.[14]

The garrison withdrew to Tripoli, where Prince Edward of England arrived soon afterwards. Edward was virtually the last great Crusader, but accomplished little before returning to England, where he soon became one of the country's greatest warrior kings, Edward I. As such, he conquered Wales and built a series of magnificent castles to enforce his control. During the rebellion of Owain Glyn Dŵr in 1401, King Henry IV appointed Henry Percy, the famous 'Hotspur', to bring the country to order. In March Hotspur issued an amnesty which applied to all rebels with the exception of Owain and his cousins Rhys and Gwilym, sons of Tudur ap Gronw of Penmynydd (forefather of King Henry VII). Most of the country was mightily relieved and agreed to pay all the usual taxes.[15] But the Tudurs knew that they needed a bargaining chip if they were to lift the dire threat hanging over them. They coolly decided to capture Edward's great castle at Conwy.

Although the garrison amounted to just fifteen men-at-arms and sixty archers, John de Massy 'of Podyngton' (Puddington in Cheshire) had put the castle in a reasonable state of defence and it was well stocked and easily reinforced from the sea; and in any case, the Tudurs had only forty men. They needed a ruse. On Good Friday, which also happened to be 1 April – All Fools' Day – Massy and all but five of the garrison were attending *tenebrae* in the little church in the town when a carpenter appeared at the castle gate who, according to Adam of Usk's *Chronicon,* 'feigned to come for his accustomed work'. Once inside, the carpenter attacked the two guards and threw open the gate to allow Gwilym and most of the gang to rush in. The rest waited outside, ready to ambush any attempt to retake the castle. Although Hotspur arrived from Denbigh with 120 men-at-arms and 300 archers, he knew it would take a great deal more to get inside so formidable a fortress. Forced to negotiate, he duly gave the Tudur boys their pardon.[16]

Medieval armies were *ad hoc* affairs, formed for the duration of hostilities and commanded by captains whose obligations were usually feudal, and who generally regarded each other as equals whether they led fifty men or five thousand. Discipline was lacking and unit training practically non-existent. This state of affairs came to an end during the late fifteenth century, when the Swiss fought for independence and, having won it, hired themselves out as mercenaries. The result was the demise of the medieval pattern of warfare based on feudal obligation as mercenaries came subsequently to dominate European armies. Warfare had never achieved the ideals that chivalry claimed for it, but a new awareness of the possibilities of strategem, and a willingness to use it, were to mark warfare as it grew into a profession. In 1513 the Flemish defenders of Tournai painted lengths of canvas to resemble fortifications and deceive the English attackers as to the true extent of the defences – but then, the Flemish always were accomplished landscape artists.[17]

The Renaissance and the Age of Reason

The only work published during his lifetime by Niccolò Machiavelli – one of the greatest thinkers of the Renaissance – was *The Art of War*. Like that of most of his contemporaries, Machiavelli's military work was inspired by the ancients, particularly Polybius and Vegetius. It rejected the values that underpinned medieval warfare and took an entirely practical view of the subject, with victory as the sole criterion for success and an acceptance of every type of trickery as legitimate. Machiavelli described the ideal commander as one capable of constantly devising new tactics and stratagems to deceive and overpower the enemy.[18] But although this was a time when firearms were starting to appear in quantity on battlefields all over Europe, it was not gunpowder that underpinned this change in approach so much as the need to introduce discipline and training of a sort unknown in medieval armies.

Machiavelli's writing inspired Justus Lipsius, who in turn inspired Maurice of Nassau. Lipsius said that whoever could combine the troops of the day with the discipline of the Roman art of war would be able to dominate the earth, and it was the development of drill and the formation of the modern infantry company requiring professional officers and soldiers by Maurice and, later, King Gustavus Adolphus of Sweden, that formed the true basis of the military revolution that accompanied the Renaissance.[19] At the same time each introduced a higher proportion of musketeers to pikemen in their regiments, and with the invention of the bayonet at the end of the seventeenth century the role of firepower increased, so that the cavalry (and its associated chivalric ideal) was no longer master of the battlefield. Along with this transformation in the nature of warfare came a transformation in the political patterns that produced it, with the development of nation states. By the beginning of the eighteenth century most states possessed standing armies officered by professional soldiers for whom deception was a natural part of war.

Such modern concepts as coalition warfare began to appear, along with the division of warfare into the tactical, operational and strategic levels (which we might simplify as the direction of armies on the battlefield, between battlefields or between theatres of war). During the War of the Spanish Succession John Churchill, first Duke of Marlborough, provided a magnificent example of strategic deception. In the spring of 1704 the French and their Bavarian allies seemed poised to capture Vienna, the capital of the Austro-Hungarian empire, and strike a strategic blow that would end the Grand Alliance, of which Great Britain was part. With a revolt taking place in Hungary, there were only 36,000 Imperial troops under for a crossing. Tallard was partially deceived by this and delayed marching on Ulm while awaiting new instructions from Versailles. Instead, Marlborough was able to cross two major obstacles, the rivers Main and Neckar, and then swing away from the Rhine towards the Danube. He only informed the Dutch of his true intentions on 6 June. As Villeroi had been shadowing Marlborough, the Dutch remained safe from an offensive and Marlborough promised to return immediately in barges along the Rhine at eighty miles a day should it prove necessary. As a result, the States-General voted him their full support on 10 June and agreed to release the Danish contingent of 10,000 men as a reinforcement.[20] It was a truly brilliant feat, covering 250 miles in five weeks with only a tiny loss by the wayside, the result of foresight, superb planning and, in an age when security was practically unheard of, secrecy. The campaign culminated in the decisive defeat of Tallard at the Battle of Blenheim and the removal of the threat to Vienna.

In the days when information could only be passed as fast as a horseman could ride, and when armies could expect to march at little more than ten miles a day, the opportunities for deception on such a scale were very rare. Marlborough had not only to plan for such contingencies as the issue to each man of new shoes at Heidelberg, but to make all the necessary diplomatic arrangements with the various German princes through whose territory he had to pass, organizing credit with bankers and the laying-in of provisions. These arrangements could not be kept secret from the French, but what could be kept secret was

Do Modern complexities More this From possible

the true intention behind them and this formed the basis for the deception[21]

At the start of the war in 1701, the other great general of the age, Prince Eugène of Savoy-Carignan, demonstrated similar skill at what would now be called the operational level. Following a meeting of the Austrian war council, Emperor Leopold gave orders for the Habsburg army to enter Milan, but there were to be long delays before they could begin. Meanwhile, in February 1701, the French were permitted into Savoy and King Louis XIV sent forces to strengthen the French garrison of Milan and to occupy the famous fortresses of the 'Quadrilateral' – Verona and Legnago on the River Adige, and Peschiara and Mantua on the River Mincio – control of which ensured strategic control of Italy. The Duke of Mantua allowed the French to assume control of the Po valley under Maréchal Catinat, so that Eugène's first problem would be simply to get into Italy. With the French in occupation from Savoy to the borders of Venice and the passes blocked from the Tyrol into Lombardy, Catinat boasted that in order to enter the country the 'Imperial army would have to grow wings'.

Eugène commanded a force of 30,000 men assembled at Rovereto in South Tyrol. 'Let us only start marching and we will soon find allies,' he boldly declared; but finding allies was the least of his problems, given that the French outnumbered him by at least 10,000 men and blocked the gorge of the Adige leading from Rovereto to Verona, the only apparent approach route into Italy. According to local legend, neither cart nor horse had been able to reach the plain by any other route, so savage were the

Pays terrain
d e l m i d
Impassable — Logistics

mountains around about, so it seemed that Catinat's boast *Necessity*
was no idle one. But Eugène understood that the legend
also served to provide a cover plan, and simulating preparations for a frontal assault on Catinat, he chose instead to
take his troops over the mountain tops eastward toward
Vicenza, even though this would infringe Venetian neutrality. Hundreds of Tyrolean peasants were conscripted
to shovel away the snow and cut paths through the wild
Terragnolo and Fredda valleys before the troops could
begin the march on 26 May. Fifteen pairs of oxen were
harnessed to each gun and a total of 6,000 horse and
16,000 infantry scrambled over Monte Baldo into Italy. So
effective was his deception that as late as 30 May Catinat
was issuing warnings of an attack from the north along
the Adige. It was a truly remarkable feat and immediately
captured the public imagination. Eugène was compared
to Hannibal and his name became forever linked to the
region: a mountain stream from which he drank is known
to this day as the Fontana del Principio Eugenio. Catinat
was taken completely by surprise and never regained the
initiative.[22] *— Dolittle Raiders — similar*

Marlborough also went on to create clever deceptions at the operational and tactical levels, and managed to repeat one particular trick on two opponents. In
Flanders in 1705 Maréchal Villeroi was defending a formidable defensive position called the Lines of Brabant.
On the evening of 17 July Marlborough's engineers built
a series of twenty pontoon bridges across the stream of
the Mehaigne, suggesting a move to the south to join up
with the Dutch, who at the same time advanced towards

troops were formed into four columns and moved quietly away by 2100 hours, leaving their camp fires burning
behind them. By the time Villars realized what was afoot
it was too late. When Marlborough received a report that
Arleux and the lines behind it were deserted, he passed
the news along the column and asked it to make an extra
effort. The soldiers responded magnificently: the 18th
Regiment of Foot completed thirty-nine miles in eighteen
hours, and by 0800 hours on 5 August the Duke and his
cavalry advance guard were pouring through the lines
near Arleux. Villars was forced to retire to Cambrai.[24]

Such a manœuvre, where a commander makes a
show without intending actually to engage the enemy, is
known as a demonstration. A show which does engage the
enemy with a portion of one's force is known as a feint,
and was a favourite ruse of the Francophile philosopher
King Frederick II of Prussia (Frederick the Great). 'You
reap greater benefit from the skin of a fox than from the
hide of a lion,' he wrote, and went on to describe how 'we
endeavour to conceal the real plan and to create an illusion
for the benefit of the enemy by feigning views we do not
hold.'[25] Frederick's voluminous writings include his *Secret
Instructions* to his generals and the *Military Testaments* of
1752 and 1768. Although he never synthesized his ideas
into a single treatise, these and other works give an insight
into his thoughts at various times. Since 'a ruse might succeed where brute force might fail,' he made frequent use
of double agents, planted messages, showy concentrations
of troops or transport, or deceptive arrangements of his
forces in camp.[26] While not an innovator in the fashion of

Maurice or Gustavus Adolphus, Frederick did devise the Attack in Oblique Order. This was designed to maximize the effectiveness of Prussia's numerically inferior armies by feinting against one part of the enemy's line before concentrating by rapid manœuvre to roll it up from the flank, a tactic used most notably at the Battle of Leuthen in 1757.

Following defeat by the Austrians at Kolin on 18 June, which enabled them to relieve Prague, Frederick was forced onto the defensive. After defeating the French at Rossbach on 5 November he rushed his small army of 36,000 men back to Silesia, determined to attack the combined 70,000-strong Austro-Russian army commanded by Prince Charles of Lorraine and Marschall Leopold von Daun, which was blocking the road to Breslau. With only half the numbers of his opponents it was a bold move indeed; but Frederick felt that boldness aided by deception would make up for the disparity.[7] After rising at 0400 hours, the army was soon on the march in two great wings of infantry flanked by cavalry with a powerful advance guard to the fore. By a stroke of fortune the ground over which the battle was fought was the Prussian army's peacetime training area. Near the village of Borne, Austrian outposts were quickly driven in and Frederick made a reconnaissance. The Austrian right wing was anchored on an oak forest, but the left fell short of Lake Schweidnitzer-Wasser. Most importantly, he could see that the high ground of the Schleier-Berg and the Sophien-Berg offered a covered approach towards the Austrian left at Sagschütz. He therefore made a deployment as if to attack directly to his front, convincing Charles that he would hit the Austrian

[margin handwriting: From weakness]

right and prompting him to bring forward nine battalions from the reserve to the area of Nippern, well over an hour's march from Sagschütz. Meanwhile, the marching Prussian columns had disappeared from view, thanks to Frederick's intimate knowledge of the terrain.

As the main body moved to assault the Austrian left, the advance guard continued forward in a feint towards the right-centre. Shortly after noon the main body was in position to assault from the south through the village of Leuthen, heavily supported by artillery. At first, Charles sought to send individual battalions to meet this new threat, but with his cavalry driven from the field, he was forced to realign his entire defence to face south. At about 1530 hours the Prussians opened a concerted attack against this new line, taken in the flank by Prussian artillery fire. Leuthen fell after thirty minutes and, with the light rapidly fading, the Austrians fell back in total disorder, which quickly turned to rout. The Prussians lost a little over 6,000 men but they inflicted 22,000 casualties on the Austrians (including around 12,000 prisoners). It was probably the greatest victory of the century.[28]

Deception can make up for a lot of disadvantages.

THE DUAL REVOLUTION

During the late eighteenth century the world was once more transformed by revolution, both political and industrial. This led to prolonged warfare between France and much of the rest of Europe, during which time the British Army won its only battle honour for service on home soil, and its most bizarre: Fishguard, 1797. It belongs to the

Pembroke (Castlemartin) Yeomanry and, if more boozy than bloody, it represents a minor masterpiece of bluff over brute force and remains a tribute to Welsh pluck.[29] Theobald Wolfe Tone, founder of the Society of United Irishmen, arrived in France in early 1796 to seek aid to establish an Irish republic. In Paris he met another dashing young man of action, Lazare Hoche, commander of the Armée des Côtes de l'Océan. Hoche envisaged a *coup de main* against the Cornish coast by 1,600 French regulars and a second landing in Wales with the aim of establishing a peasant uprising in Britain. He was in the process of putting these modest proposals into effect when word arrived from the governing Directory that something rather more grand was being planned. These expeditions were to become subsidiary diversions to the main effort of putting 15,000 men ashore to assist in the liberation of Ireland.

The expedition fell foul of a rising gale off Bantry Bay, and for a fortnight Hoche, Tone and their army were borne about on the back of an Atlantic gale, which forced them to abandon the attempt. Immediately afterwards, the plans for raids on Cornwall and Wales were dusted off again; the Cornwall scheme was then dropped, but Tone had spent some of his time in translating orders for the American leader of the expedition aimed at Wales, William Tate. Tate's orders were to land within five miles of Bristol at dusk. Having destroyed what was then England's second city, he was to cross over to the right bank of the River Taff and march on Chester and Liverpool. His ragtail 'army' was assembled from the dregs of the prisons, pressed émigrés, and a few released prisoners of war who evidently did not know what they had volunteered for. They were issued British uniforms captured at Quiberon and dyed deep brown, which earned them the title Légion Noire.

After they had raided Ilfracombe conditions simply would not permit the passage up the Bristol Channel. Tate then declared Cardigan Bay his alternative objective, and course was duly set. The squadron was sighted on the morning of Wednesday 22 February off North Bishop Rock.* Shortly afterwards, Tate's men seized a local man, John Owen of Pencaer, from his sloop *Britannia* and quizzed him as to the defences of the area. Helped by some brandy, he greatly exaggerated the defenders' numbers, but his estimate still amounted to less than half that of the invaders. Soon seventeen boatloads of uniformed cutthroats and brigands descended upon as peaceful a spot as exists in Western Europe. Forty-seven barrels of gunpowder and 2,000 stands of arms for the proposed uprising were also landed.

To defend the area, John Campbell, Lord Cawdor, proceeded to assume command of the 400 assorted men, including the Castlemartin Yeomanry, assembled at Haverfordwest. These then set off towards Fishguard,

---

* One of the ships rounded Pen Anglas into Fishguard Bay to be greeted by the 9-pounder ordnance of the fort. This was firing the alarm to summon the local volunteers, and the ship withdrew, unaware that the fort's eight 9-pounder guns had only three rounds of ammunition and sixteen cartridges between them. It was the townsfolk's responsibility for the supply but they had never taken the threat of invasion seriously (E. H. Stuart Jones, *The Last Invasion of Britain,* pp.72-4).

while most of the French troops were busy looting the surrounding countryside and getting into skirmishes. One local woman, Jemima Nicholas, a 47-year-old cobbler, marched resolutely out to Llanwnda armed with a pitchfork and promptly rounded up twelve Frenchmen, whom she brought into town before departing to look for more. Cawdor's force arrived as the evening drew on, and planned an immediate attack. But the fight never developed as they could not manœuvre their improvised artillery through the narrow lines, and they decided to wait for morning.

Dismayed by what he saw, Tate decided to seek terms. At eight o'clock he sent his second-in-command, the former Baron de Rochemure, and his English-speaking ADC to deliver a missive:

> Sir,
> The Circumstances under which the Body of French troops under my Command were landed at this place renders it unnecessary to attempt any military operations, as they would tend only to Bloodshed and Pillage. The Officers of the whole Corps have therefore intimated to me their desire of entering into a Negociation upon Principles of Humanity for a surrender. If you are influenced by similar Considerations you may signify the same by the bearer and, in the mean Time, Hostilities shall cease.[30]

Cawdor must have greeted this development with delight and may have also been tempted when, shortly afterwards, de Rochemure announced that the only detail requiring agreement was the repatriation of the French at the British government's expense. But Cawdor refused even to contemplate this and, cleverly disguising his weakness, offered the following grandiloquent reply:

> Sir,
> The Superiority of the Force under my command, which is hourly increasing, must prevent my treating upon any Terms short of your surrendering your whole Force Prisoners of War. I enter fully into your Wish of preventing an unnecessary Effusion of Blood, which your speedy Surrender can alone prevent, and which will entitle you to that Consideration it is ever the Wish of British Troops to show an Enemy whose numbers are inferior.[31]

This was an outrageous bluff but it prompted Tate to communicate the following morning that he would surrender under any terms and articles were duly prepared.

Tate must have seen that he outnumbered Cawdor's rag-tag army (many were sailors and at least a fifth were volunteer civilians). Yet a procès-verbal drawn up by his officers on 25 February and signed by him spoke of the British coming at them 'with troops of the line to the number of several thousand'.[32] Thousands of people gathered to witness the Légion Noire lay down its arms on Goodwick Sands, including women clad in the then fashionable scarlet mantles and low-crowned round felt hats. These may have appeared to the French like British Army redcoats at a distance. The official French historian Captain Desbrière refers to 'un rassemblement de femmes galloises', and a letter from John and Mary Mathias to their sister in service in Swansea describes 'near four hundard Women in Red Flanes and Squier Cambel went to ask them were they to fight and they said they were'. It is easy to picture a crowd of women coming to watch the proceedings being asked by 'Squier Cambel' if they had come to fight, and being eager to take a hand. The deception may not have been intentional, but its effect was the same.[33]

In invading Britain, Tate achieved one thing that always eluded Napoleon Bonaparte, one of the world's greatest generals and a master of deception.[34] Napoleon was a voracious reader, but he left no body of writing to students of his military art. Instead, his art was handed down by his actions and the reports of others. Like Marlborough and Frederick, Napoleon was not so much a military innovator as a skilled manipulator of the tools available. Although he disliked categorizing his methods, he operated three broad types of manoeuvre.[35] The *manœuvre sur les derrières* (or strategical envelopment) was demonstrated early in Napoleon's career by the Manœuvre of Lodi in 1796. Here he was faced with a crossing of the River Po, which was contested by the Austrian general Johann Beaulieu. Napoleon's plan was to distract Beaulieu while he himself moved eastwards to Piacenza; there he would establish a bridgehead from which, if he could capture the crossings over the River Adda which flows south into the Po, he would threaten the Austrian rear. This was achieved by mounting demonstrations that appeared to presage a crossing in the area of Valenza while a chosen force marched hard for the real objective, thus succeeding in getting behind the enemy and threatening to cut it off.[36]

When Napoleon came to prominence, the French Army had been as imbued with revolutionary fervour as the rest of the country. It burnt with a patriotic zeal previously unseen, and the *levée en masse* created the first mass conscript armies. Marlborough and Frederick had been forced to keep their much smaller armies together, largely for logistic reasons (troops were not allowed to go foraging for supplies for fear of desertion, so all supplies had to be carried in large wagon trains), but the French revolutionary army was capable of operating with greater freedom of action and less reliance on depots than previously. The patriotism of its soldiers meant they could be trusted to forage for themselves, and since they were fighting largely on foreign soil, the burden did not fall on France. Napoleon realized that the eighteenth-century pattern of siege warfare had led to endless logistical problems, and since the large armies available to him meant he could screen off fortresses and not worry about sieges, the logistic apparatus that previously limited an army's freedom of manoeuvre could be dispensed with.[37]

Given these factors, Napoleon created his greatest innovation – the army corps. By organizing the Grande Armée into all-arms groupings, each capable of independent action and of looking after itself until support arrived, he was able to advance on a wide frontage in a manner not seen since the Mongols, thus enabling him to

- DECENTRALIZED EXECUTION

Gives more DECEPtive
Capabilities — SECRITY/Secrecy

cloak his intentions and main effort. As a result, his march to the Danube in 1805 marked the transition between eighteenth- and nineteenth-century warfare. He was able to advance with 200,000 men on a frontage of nearly 200 miles, reducing to 70 miles when he reached the river, and to trap an Austrian force of 27,000 men at Ulm under the totally bemused Karl Freiherr von Mack, as well as taking a further 30,000 prisoners in the days that followed. Unable to pull a similar trick on the Russians who were coming up to support the Austrians, Napoleon skilfully feigned weakness, and the combined Austro-Russian army advanced to attack him at Austerlitz. From the commanding Pratzen heights the Austro-Russian force looked down on Napoleon's apparently weak right wing, and moved to encircle it in four great columns totalling 40,000 men. Having thus lured the allies out of position, Napoleon with perfect timing unleashed previously concealed troops into the gap created in the centre of the Austro-Russian line, and achieved his greatest tactical victory.[38] Thereafter, his decision to invade Russia notwithstanding, he proved perhaps more skilled as a strategist than as a tactician.[39]

Warfare and revolution continued throughout Europe for the remainder of the nineteenth century, but while Britain and France in particular also took the opportunity that industrialization presented to extend their empires, another truly great commander welded a tribe of perhaps 1,500 into a mighty nation that in due course would humble the greatest empire of all. King Shaka of the Zulus developed a revolutionary war machine based on the stabbing assegai and a regimental system that swept all before him. He was also a great deceiver and delighted in luring the enemy into positions favourable to himself.[40] In his first full battle against the Butelezi in 1816 he bunched his regiments at the outset and had his men carry their shields on edge to make his force appear small. When the horns of his famous bull's head formation raced out, the warriors turned their shields outwards making the army instantly appear double its original size.[41] At the Battle of Gqokli Hill in 1823 Shaka faced a far greater force of the Ndwandwe. He sent the Zulu cattle off with a small escort, but deliberately left the herd visible in order to draw off a portion of the enemy in pursuit; then, abandoning his usual tactics, he occupied a position on top of the hill, where he concealed his reserve in a deep depression. After their initial assaults had been resisted, the Ndwandwe formed a column, intending to drive Shaka off the hill onto a cordon at the bottom. Instead, they were in turn surrounded by the hidden reserve, a combination of skill and cunning that brought off Shaka's greatest victory.[42]

— position/posture of centres of gravity.

THE AMERICAN CIVIL WAR

Western technological development transformed warfare and eventually swept away the Zulus' world. Although many of the developments for which it is famed – the use of railways, telegraph and ironclad steamships – were not in fact new, the American Civil War is nevertheless often referred to as the first 'modern war'. In terms of scale – with its mass armies, mass production and mass casualties

– it certainly did represent modernity, but it was fought using largely Napoleonic tactical methods.

Following a disastrous opening to the war at First Bull Run (or Manassas Junction) in 1861, the Union appointed Major-General George B. McClellan as general-in-chief. His urgent task was to reorganize and train the Army of the Potomac, both for the defence of Washington DC and for future offensive operations with a view to capturing the Confederate capital of Richmond, Virginia. 'I can do it all', McClellan assured President Abraham Lincoln. McClellan was known as 'the Young Napoleon' and affectionately as 'Little Mac' by his troops. He was an able administrator and trainer, but lacked resolution in the face of the enemy. At this time the Union intelligence service was run by the Pinkerton Detective Agency, founded by the ex-Glaswegian Allan Pinkerton, which later became renowned throughout the West. As a military intelligence bureau, however, it was hopelessly inadequate and provided wildly exaggerated reports of rebel strength in the area immediately south of Washington. Despite ample evidence to the contrary, Pinkerton reported Confederate forces as totalling 270,000 men, with 150,000 within striking distance of Washington. Little Mac refused to move until he had 270,000 men of his own.

Then in September rebel pickets were driven surprisingly easily from a position they had occupied within a few miles of Washington, revealing that the guns McClellan's spies had assured him were trained on the capital were nothing more than stripped logs, painted black with wagon wheels tacked onto the side; one scornful reporter christened them 'Quaker Guns'.[43] Lincoln became so frustrated with Little Mac's lack of resolution that when the latter was ill early in 1862, Lincoln told a White House war council that 'if General McClellan does not want to use the Army, I would like to *borrow* it for a time'.[44] Eventually however, McClellan was persuaded to take the offensive, albeit not via the direct route (which he remained convinced was strongly defended) but by a landing on the York-James peninsula and approaching Richmond from the south-east. The Confederates in front of Washington then abandoned their position to reveal an entire battery of Quaker guns at Centerville.

On the peninsula Little Mac's army, which totalled over 120,000 men, was initially faced by just 8,000 Confederates under John Bankhead Magruder, a lover of amateur theatrics known as 'Prince John' because of his lavish parties, fancy dress uniforms and pomposity (he even affected a 'Horse Guards' lisp). Friends recounted how he had tried to impress visiting British officers with his dinner and wines and displayed surprise when asked how much American officers earned, saying he had no idea and would have to ask his servant. This, like his entire lifestyle, was in fact a grand bluff; he had no independent income at all.[45] Now he set to bluffing with a will.

Having anchored off Fort Monroe on 2 April 1862, Little Mac was initially filled with optimism but soon became despondent when the roads proved far worse than expected, slowing his baggage and artillery. Although Magruder had built his defence works with great energy he had a thirteen-mile line to defend and there were simply not enough guns to cover it: he had been able to secure just fifteen, including light field pieces, and had barely sixty rounds for each. Therefore he made up the numbers with Quaker guns, hoping to replace them all with real ones in due course, but McClellan had arrived before he had a chance. So Magruder mixed Quaker guns with real ones along the line, hoping this would prove sufficient to delay the advancing enemy just enough, which with the cautious McClellan proved the case. With 67,000 men immediately to hand, McClellan could have brushed Magruder aside, but to add colour to the deception, Prince John conspicuously moved his handful of units about and ordered his bandsmen to play loudly after dark, while he himself rode ostentatiously about with a colourful following of staff officers. One battalion was sent to march along a road that was heavily wooded, except for a single gap in plain view of the Union lines. In an endless circle through the same clearing they swept past in seemingly endless array. '[We] have been travelling most of the day, seeming with no other view than to show ourselves to the enemy at as many different points of the line as possible,' wrote an Alabama corporal, 'I am pretty tired.'[46]

It worked all too easily. Little Mac halted his infantry when it could have walked through the Confederate position at any point it chose and ordered his artillery to begin probing the defences. As early as 7 April he was telegraphing Washington to whine: 'General J. E. Johnston arrived in Yorktown yesterday with strong reinforcements. It seems clear that I shall have the whole force of the enemy on my hands, probably not less than 100,000 men and possibly more.' He believed therefore that his own force was 'possibly less than that of the enemy'. No attack could succeed and, 'were I in possession of their entrenchments and assailed by double my numbers I should have no fear as to the result.' Despite intelligence reports that the enemy had no more than 15,000 men (which McClellan acknowledged as early as 3 April), the Young Napoleon believed that nobody, still less a professional soldier, would try to hold so precarious a line with so few.[47] On 5 April McClellan declared, 'I cannot turn Yorktown without a battle, in which I must use heavy artillery and go through the preliminary operations of a siege.' In fact, by the 11th Magruder's force still amounted to just 34,000 men and Johnston did not even reach Richmond until the 12th. When the Confederates eventually retired on the night of 3 May, just as Federal siege preparations were being finalized, their forces amounted to only 56,000 men. McClellan, who had been deeply impressed by his visit as an observer to the siege works of Sebastopol in the Crimea seven years previously, was probably more impressed with the works facing him than the apparent size of the garrison.[48] Nevertheless, the diarist Mary Chesnut recorded that 'it was a wonderful thing how [Magruder] played his ten thousand before McClellan like fireflies and utterly deluded him – keeping down there ever so long.'[49]

Another Confederate general put on a command performance in May 1862. At Corinth, Mississippi, following the Battle of Shiloh, Major-General Pierre G. T. Beauregard sent 'deserters' to the Union lines with carefully rehearsed stories about his 'offensive' plans together with cavalry raids to spread panic and rumour. But he knew he could not hold the town if it came to a siege and decided that, in order to save his army, a retreat was necessary. Keeping his plans a secret from all but those who strictly needed to know, he arranged to evacuate the wounded, send on baggage and even remove the signposts beyond the town to hinder any pursuit. Meanwhile, with all his bands playing, a regiment was kept cheering the trains that arrived to take away his wounded, to convey the impression that reinforcements were arriving.

When the time came to tell the front-line soldiers that they were to withdraw, they were happy to join in the fun. They stole out of their trenches that night, leaving drummer boys with wood supplies to tend their fires and beat reveille in the morning, together with a single band to play at various points and a detachment to continue cheering the single train of empty cars that rattled back and forth in and out of the station all night. At 0120 hours that morning the Union commander, Major-General John Pope, sent word to his superiors that 'the enemy is reinforcing heavily, by trains, in my front and on my left ... I have no doubt, from all appearances, that I shall be attacked in heavy force at daylight.'[50] Instead, when daylight came, according to Brigadier-General Lew Wallace, the Union troops found 'not a sick prisoner, not a rusty bayonet, not a bite of bacon – nothing but an empty town and some Quaker guns.'[51] Worse, the dummy guns were served with dummy gunners fashioned from straw and old uniforms.

Nathan Bedford Forrest, 'the Wizard of the Saddle', was described by William Tecumseh Sherman as 'the most remarkable man our Civil War produced on either side', although this did not prevent Sherman ordering that Forrest be 'hunted down and killed if it cost ten thousand lives and bankrupts the federal treasury'.[52] Forrest's instinctive, brilliant command of cavalry included a flair for deception: he consistently managed to exaggerate his strength by a considerable margin. When he crossed the Tennessee River near Clifton on 17 December 1862, he needed to complete his task before the Union had time to concentrate forces for his destruction. Having captured some Union civilians, he drilled his men as infantry in their presence before allowing the civilians to escape, and in this way spread the rumour that his command included a large body of infantry. By the same token his men always carried a number of kettledrums which they kept beating to further reinforce the impression that there were infantry with him.[53] *Using CoIPS as TaIl/SALUTE*

In April 1863 Forrest was given the task of defeating a Union raid into Alabama by Colonel Abel D. Streight. When Forrest finally cornered Streight and demanded his surrender, Forrest claimed to have a column of fresh troops at hand. 'I have enough men to run straight over you,' he said.[54] Streight refused even to contemplate laying down his arms unless Forrest could prove this was so, but Forrest would not show his hand. Meanwhile, as previously instructed, Forrest's artillery commander repeatedly brought his two guns over a rise in the road, into cover and round again, which Streight could observe over Forrest's shoulder. 'Name of God', cried Streight at last, 'how many guns have you got? There's fifteen I've counted

reinforcements. These arrived shortly afterwards and were also compelled to surrender.[56] *- Forrest - Deception about strength*

B-P AT MAFEKING

Another man famed for his love of amateur theatrics was Colonel Robert Stephenson Smyth Baden-Powell. B-P (as he was universally known) was *the* hero of the Second Anglo-Boer War and arguably saved South Africa for the British.[57] Although the siege of Mafeking lost its strategic importance within a few weeks of its start, the British public, rocked by the disasters in December 1899 collectively known as Black Week, combined with the ignominy of thousands of British regulars being cooped up in the sieges of Kimberley and Ladysmith, were enthralled by B-P and Mafeking. Here, it seemed, a bunch of amateurs under an obscure colonel was making fools of the Boers. In fact, these were precisely B-P's instructions. 'As an actual feat of arms', he wrote later, it 'was largely a piece of bluff, but bluff which was justified by the special circumstances'.[58] Meanwhile, his sardonic dispatches – 'One or two small field guns are shelling the town. Nobody cares' – further endeared him to the British public.

In June 1899, with war in South Africa approaching, B-P was sent with the grand-sounding position of Commander-in-Chief, North-West Frontier Forces, to raise two battalions of mounted infantry and to co-ordinate the police forces of the region. Furthermore, he had secret instructions in case of war to raid Transvaal and draw off as many Boers as possible from the vulnerable Cape

Colony and Natal. In fact, the tomfoolery and bluff with which he made his reputation were part of his orders from the War Office.[59] But at the end of September, with hostilities imminent, the Cape administration forbade him access to the town, forcing him to bluff his way in past the authorities: 'I got permission from the Cape to place an armed guard in Mafeking to protect the stores; but as the strength of the guard was not stipulated I moved the whole [Protectorate] regiment into the place without delay.'[60] War was declared by President Paul Kruger of Transvaal on 11 October and the Boers swallowed the bait of Mafeking whole, immediately investing it with around 8,000 men.

The scratch garrison amounted to just 48 officers and 1,183 men; not all had modern weapons and there was no modern artillery. Unable to raid Transvaal, B-P set about achieving his aim from within the confines of the town, but there was no way of preventing the Boers from cutting Mafeking off from supply and reinforcement, and even the most sanguine estimate put relief six weeks away B-P was almost alone in thinking that even the initial assault could be turned back. He wanted to convince Commandant-General Piet Cronje that the toothless lion of a town had in fact got sharp claws. He immediately set up a chain of outposts on a five-and-a-half-mile perimeter, a wide area for so small a force but one at such distance that he hoped would prevent an overwhelming rush and which included the 6,000 natives in the Baralong township. One fort was built a mile and half to the west of the railway with mounds of earth, sandbags and two outsized flagpoles clearly marking it as his own headquarters. In due course it drew much enemy fire, as he hoped it would: it was a dummy. Thornbush was woven into *zareba* instead of barbed wire and suitable houses were loopholed and prepared. Trenches were dug and a breastwork made of stone at the old fort on Cannon Kopje. All these positions were then linked by telephone to Dixon's Hotel, where B-P set up his real headquarters with a lookout position that gave him a fine view of the area. Food did not appear to be a problem since few people expected a siege to last long. The Dutch among the population positively gloated, but the numerous Boer agents in town would prove invaluable to its defenders.

Work on the defences included strings of natives continuously carrying boxes gingerly about town, telling anyone who asked that they must not be dropped. 'Minefields' soon began appearing all round town with prominent warning signs in Dutch and English and given wider publicity by official announcements. B-P made a very public show of test-firing one. 'With everyone safe indoors,' he later wrote

Major Panzera and I went out and stuck a stick of dynamite into an ant-bear hole. We lit a fuse and ran and took cover until the thing went off, and it did with a splendid roar and a vast cloud of dust. Out of the dust emerged a man with a bike who happened to be passing, and he pedalled off as hard as he could for the Transvaal, eight miles away, where he no doubt told how by merely riding along the road he had hit off a murderous mine. The boxes were filled with nothing more dangerous than sand![61]

Slowly the ring was tightened but for some days, apart from probing actions, the Boers did little. It was apparent that Cronje's aim was a morale-boosting bloodless victory The Boers were in control. Or so they thought. The perimeter was never entirely closed, enabling information at least to pass both ways. B-P kept another 1,200 Boers idly watching the southern stretch of the Bechuanaland border. Weeks earlier he had written to an old English acquaintance who ran a farm just inside Transvaal to the north, warning him of the approach of a 'Third Column'. B-P knew, however, that the man was dead and that consequently the letter would be opened and its contents passed to the enemy. The 'Third Column' only ever existed in the minds of the Boers. Lieutenant-Colonel Herbert Plumer's 500 men of the Rhodesia Regiment did exist, though, and they made a great show of themselves to occupy another 2,000 Boers along the Limpopo River and prompted urgent telegrams from Kruger asking 'where is Plumer?' and urging the Boers to 'watch Plumer at all costs'.[62]

The Boer reluctance to attack now suited B-P, at least during the day. At night lay obvious danger until one evening, shortly after sunset, an intense white light shone out from one of his outposts. Shortly thereafter (or possibly at the same time for the Boers, were notoriously bad timekeepers) another appeared at a different location and then another, until it seemed to the Boers that they faced not only mines but also many searchlights. Sergeant Moffat, in charge of the garrison signallers, assisted by Mr Walker of the South African Acetylene Gas Company, had rigged a contraption by soldering biscuit tins together to form a rectangular cone with a gleaming interior and an acetylene torch through the bottom, all attached to a long pole. They quickly moved this from fort to fort, which greatly impressed the Boers when shone in their general direction.[63] Another contrivance made from biscuit tins was a megaphone that could clearly convey words of command more than 500 yards. This was deployed in forward positions to broadcast carefully rehearsed conversations in which B-P played the leading role, issuing orders to notional subordinates to prepare to attack with fixed bayonets. The Boers were scared witless by this weapon and its threat would draw a wild barrage of musketry, disclosing their own positions and assisting the watching British snipers.[64] Such efforts conformed to B-P's general instructions to the garrison to 'bluff the enemy ... as much as you like'.[65]

B-P published some correspondence between himself and Cronje in which Cronje admitted that Mafeking could not be taken by assault and B-P again referred to his mines. Seven times through October and November B-P sent out sorties to give the Boers what he called 'kicks'. Although expensive in casualties, they had the desired effect of keeping the attackers on the back foot. Greater resolution would surely have swept the defenders away but Kruger, although he instructed Cronje to 'make an end of it', ruled out attacks likely to result in more than fifty casualties.

When news arrived of the investment of Ladysmith, early relief was clearly impossible and led on 17 November to the introduction of rationing. Thankfully, this indication of the future was overshadowed the following day by the news of the departure of Cronje and most of his men. B-P and his two regiments of 'loafers' staffed by a dozen Imperial officers had produced an important strategic victory, distracting Cronje and a quarter of the Transvaal's force, including a score of modern guns, for over a month. General J. P. Snyman, who replaced Cronje, made no attempt to take Mafeking by assault and both sides settled down for a long wait. About seventy shells a day were fired into Mafeking throughout November and December, to little effect. B-P continued to supervise the improvement of the defences and, on noticing the Boers stepping high over barbed wire attached to wooden pickets, set out pickets of his own which his men ostentatiously stepped over – they were tied with string.[66] With alertness and guile he maintained the defence against what remained considerable odds. After 217 days Snyman finally departed. B-P had displayed the 'audacity and wariness' recommended in his original War Office instructions together with an utterly ruthless will to win which, combined with the irresolution of both Cronje and Snyman, had brought him a deserved victory.

Finally, the beginning of the twentieth century provides one particularly cold-blooded and callous example of a general deceiving his own side. The perpetrator was the German general Erich von Falkenhayn, at the Battle of Verdun in 1916. Following their defeat at the hands of

Prussia in 1870 the French built a string of fortifications between the Swiss and Belgian frontiers while they nursed their plans for revenge. The principal strong- point of this system, already fortified by the Romans and later by Vauban, was at Verdun. In December 1915 Falkenhayn, who had been appointed Chief of the General Staff, addressed a memorandum to the Kaiser in which he argued in a convoluted fashion that Germany's principal enemy was Britain and that the best way to defeat her was to knock the French Army out of the war. He went on to describe objectives that the French would throw in every available man to retain, and as a result of which their forces would 'bleed to death'. If this were not chilling enough, the plan that he proposed involved the German Fifth Army launching 'an offensive in *the direction* of Verdun'; yet Crown Prince Rupprecht of Bavaria, who commanded the Fifth Army, never saw the original memorandum and in due course issued orders '*to capture* the fortress of Verdun by precipitate methods'. Falkenhayn approved this order, even though he himself had no such intention, since the capture of Verdun would remove the carrot that was designed to draw the French into the mincer. Apparently, he calculated that the Fifth Army would fight better if they thought they were to capture the fortress rather than engaging in a battle of attrition. Falkenhayn went on to promise the Crown Prince that adequate reserves would be available but deliberately withheld them. Of all the deceptions wrought over the centuries, few examples are more cynical than this.[67]

CHAPTER TWO

THE INFORMATION BATTLE

'All the business of war, and indeed the business of life, is to endeavour to find out what you don't know by what you do; that's what I call "guessing what was at the other side of the hill"'.

The Duke of Wellington

THE INTELLIGENCE PROCESS

AN OLD FRENCH book on bridge supposedly started with the words: 'Rule 1. Always try to see your opponent's cards.' Naturally, a general who knows his opponent's intentions has a similar advantage. The deceiver knows that the enemy also wants to see his cards, and his purpose is to display false ones. A knowledge of the enemy's intelligence capabilities and weaknesses will facilitate feeding him false information and help ensure that he accepts it. If the enemy has a predilection for particular sources of information, deception planning can be tailored accordingly: the deceiver must know where to put the right cards so that they will be seen, noticed and, most importantly, acted upon.

Intelligence has long been associated in many people's minds with espionage, thanks largely to spy fiction. But for centuries the word intelligence meant news of any sort, and newspapers would head their columns 'Foreign Intelligence' or 'Domestic Intelligence'. In military parlance it is important to distinguish between information and intelligence. The former might be a bald fact such as 'the enemy has arrived at the river,' while the latter concerns the significance of such a fact: if, for example, the enemy's bridging pontoons are in the next county he will be unable to cross the river for some time. Intelligence is *Intel* thus the process of recording new information and relat- *defined* ing it to what is already known, determining the credibility of the source and then analysing it. Information therefore only becomes intelligence after it has been processed. The business of collecting information – about the enemy (preferably without his knowledge) but also about physical conditions, local supply sources, the population or any other factor that might affect operations – is reconnaissance. Surveillance involves the systematic observation of selected areas and is an inherent part of reconnaissance. It also helps to provide security, which aims to ensure freedom of action and prevent or restrict the enemy's reconnaissance and surveillance activities. However, none of this is as easy as it sounds, especially in the face of the enemy's own reconnaissance and surveillance activities.

The specific purpose of military intelligence is to forecast what the enemy will do, where and when he will do it, how and in what strength. To be of any use, this must be disseminated to decision makers as quickly as possible. It thus bears a certain similarity to weather forecasting, always bearing in mind that there is a distinct and important difference between an enemy's capabilities (which are relatively easy to define) and his intentions (which seldom are). This is especially true when viewed in the context of Helmuth Graf von Moltke's observation that if there are three courses of action open to the enemy, he invariably chooses the fourth. Prediction inherently involves a measure of informed guesswork, and as a result some commanders have felt their guesses to be as good as those of their staff, regarding briefings as no more than a means of bringing them up to date on what has happened rather than on what will happen. In such circumstances the intelligence staff become merely diarists and historians. This may be significant to those seeking to deceive them: attempts to deceive the Japanese in Burma during the Second World War frequently failed because of the conceit and inflexibility of Japanese commanders on the one hand (making Allied efforts to induce them to change their plans unlikely) and the low esteem in which the Japanese intelligence service was held on the other (it was woefully inefficient and frequently ignored).[1] By contrast, the accurate prediction by German intelligence that the French and British would do nothing enabled Adolf Hitler to leave just twenty-three weak divisions covering the West while he overran Poland in September 1939.

To compare the task of building an intelligence picture to that of a making a jigsaw puzzle is too simple an analogy since a jigsaw is neat and systematic, whereas a 'great part of information obtained in War is contradictory, a still greater part is false, and by far the greatest part is of doubtful character'.[2] A more useful analogy is that of painting a picture, where each stroke of the palette knife is a piece of information. Compare the painting styles of a neo-impressionist such as Georges Seurat and an abstract artist such as Jackson Pollock. Seurat's style (known as 'divisionism' or 'pointillism') is like a form of mosaic, in which colours are applied to the canvas in a series of small spots that, when viewed from a distance, reveal a clear image of people and landscape.[3] In contrast, Pollock's coloured mosaics are abstract and do not take a recognizable form. In intelligence terms this abstraction is interference or 'noise' – contradictory indicators, missing data, fast-moving events and time lags between data collection and analysis, and pure chance – all of which inhibit accurate intelligence assessment.[4] The aim of the intelligence officer is to watch the picture as it takes form and predict what it will become. But in attempting to create a misleading image the deceiver is not trying to fool the opposing intelligence officer so much as the opposing commander, a process that requires an understanding of both the opponent's intelligence processes and, as the Japanese have demonstrated, the enemy commander's attitude towards it.

The intelligence process takes the form of a simple cycle. The first action is direction: the commander must tell his staff what he needs to know so that they can allocate resources to collect information. Collection forms the second stage, and as the information comes in, it must be processed into intelligence and then disseminated to those who need it. By constantly re-evaluating what is known by what is not, the cycle continues. From a deceiver's point of view, the critical phases of the enemy's intelligence cycle are the collection and processing phases. It is towards the enemy's sources and agencies that false information must be directed, and a knowledge of what he is looking for during processing will assist in sending the 'correct' wrong information, since it is by reading 'signatures' of operation that intelligence staffs make predictions. For example, a combination of knowing what purpose a particular piece of equipment fulfils, and its relative position in the order of battle, can be used as a signature. Certain equipment, such as particular anti-aircraft systems, might be held at corps or army level, and their positions might therefore indicate either a corps headquarters or an army axis of advance. If one side is defending a river line, the location of the enemy's bridging equipment may indicate where an attempt at a crossing will be made. Each army has its own characteristics, which must be carefully studied, and knowledge of one's own characteristics immediately opens deceptive possibilities for the display of false ones.

However, intelligence was long considered the poor relation. At the start of the Second Anglo-Boer War in 1899, the Intelligence Department at the War Office had a budget of just £20,000 to cover the whole world (a quarter of which was governed by Britain, which had made quite a few enemies in the process).[5] By the time Aldous Huxley noted the distinction in the *Encyclopedia Britannica* between the separate articles on 'Intelligence,

human', 'Intelligence, animal' and 'Intelligence, __
there had long been a common perception of something
unreal about the concept of military intelligence, as
though all soldiers are idiots by nature. Folk memories
of the First World War reinforced this fallacy, obscuring
the transformation in the nature of warfare that had subsequently taken place, which in the field of intelligence
included the significant developments of electronic warfare (EW) and aerial photography. However, the importance of proper intelligence was increasingly understood
by the British and in 1940 it was formalized in the Army
with the formation of the Intelligence Corps. If in France
the Deuxième Bureau was efficient and good at its work,
in Germany intelligence became increasingly fractured
as Adolf Hitler's cronies sought to carve out little empires
for themselves. Meanwhile, both the USA and Japan dis-
regarded the importance of intelligence, with short-term
and long-term catastrophic effects respectively.

SOURCES AND AGENCIES

An intelligence source is anyone or anything from which
information can be obtained. An intelligence agency is
any organization or individual dealing in the collection
of information for intelligence use. Before the second
half of the nineteenth century intelligence organizations,
if they existed at all, were rudimentary and often relied
on one person's drive and ingenuity, very often that of the

---

* *Point Counter Point,* London, Flamingo, 1994, p.83.

— We still lack a full account of Marlborough's spies

commander himself. Time and again Marlborough used ruses and speed to conceal his intentions and to divine those of the enemy through superior intelligence activities masterminded by Cadogan.[6] Not only was Cadogan in charge of Marlborough's administrative arrangements, he was also Marlborough's chief of staff and chief of intelligence. An officer whose attention to detail transformed Marlborough's broad concepts into practicable orders, Cadogan provided security for the army's train and carried out myriad ancillary duties and special missions. He ran 'correspondents' in Mons and Lille and, disguised as a peasant, personally investigated the Lines of La Bassée before the Battle of Malplaquet. It was Cadogan to whom Marlborough turned whenever there was need for a reconnaissance or to lead an advance guard.

Almost all intelligence was derived from spies, prisoners, locals and other people (what is today referred to as human intelligence or HUMINT) or else by reconnaissance on foot or horseback, with the possible assistance of high ground or a telescope. Frederick the Great wrote that 'if you know the enemy's plans beforehand you will always be more than a match for him, even with inferior numbers,'[7] and he himself devoted much effort and imagination to gathering intelligence, especially the long-term strategic kind. A Jew, I. Sabatky, acted as Frederick's liaison with corruptible Russian officers (many of whom were in fact German) and he had at least one spy in the camp of the Austrians. Personable and resourceful young men acted as 'sleepers' in Vienna, where they melted into society and got themselves on intimate terms with the serving girls of the great ladies.

Honey dicks

The discoveries made by these young Adonises were quite incredible. Some of these gentlemen maintained liaisons with the Viennese chamber maids for a couple of years on end, and they wrote reports which contained far greater and more important disclosures than all the despatches of the envoys.[8]

However, day-to-day operational intelligence was usually lacking altogether. In this respect, Frederick's spies were of little use to him for he paid most of them poorly and then refused to believe them when they brought him bad news. (From a strictly military point of view, spies have seldom proved effective sources of information.) Frederick's staff was very small, and the myopic king himself became the eyes of the army when he rode out on reconnaissance with the advance guard or a little escort. He looked out not only for the positions of the enemy troops but also for signs such as smoke from camp fires and bakeries or for any indication that the Austrians were on the move. This was dangerous work, for it brought him within the zone of the enemy outposts.[9]

Cavalry has played a major role in reconnaissance from at least the time of Hannibal and his excellent Numidians. One of the hallmarks of Napoleon's art of war was his use of light cavalry – hussars, lancers and chasseurs. They scurried ahead of the hurrying columns forming a dense mobile screen, scientifically probing every village and emptying every postbox in their search for information about the enemy, perhaps capturing a prisoner or two or finding a handful of deserters, and listening to local gossip. From this mass of information Napoleon and his staff would at least be able to establish where the enemy was not situated, and thus build up an idea of where he might still be.¹⁰ Similarly, the Confederate general Robert E. Lee relied on the cavalry of J. E. B. Stuart for information on the whereabouts of the Army of the Potomac: the absence of Stuart for a week before and at the beginning of the Battle of Gettysburg famously deprived Lee of critical information and is often cited as a reason for the failure of that ill-fated incursion into Pennsylvania. But Lee made careful use of other sources too, including Northern newspapers, scouts, spies and friendly civilians who came through the lines. He had a highly developed intelligence procedure in which he not only tried to put himself in the other man's position, but actually to become that man.¹¹

In modern mechanized warfare ground reconnaissance continues to play a vital role in gathering information. In British parlance close reconnaissance applies to activities conducted within a few kilometres of the front line and is carried out from a unit's own resources, using foot patrols and observation posts. Medium reconnaissance is a specialist task still carried out by cavalry regiments (albeit mounted in armoured vehicles) at a distance of anything up to fifty kilometres ahead of the front line of one's own troops. Long-range reconnaissance is also a specialist task, often carried out by special forces. The British Army has traditionally relied on stealth as the means of obtaining such information and has equipped its recce units accordingly. The Russians on the other hand, and to a lesser extent the Germans, have always been happy to fight for information, using small all-arms groups to force an opponent to reveal his hand. Other specialist means of reconnaissance include sound and flash location of artillery positions, artillery location and ground surveillance radars which, with the wide variety of night viewing devices and many other specialist sensors that are available nowadays, make the task of reconnaissance and surveillance an increasingly complex one.

Two significant sources throughout history have been captured enemy documents and prisoners of war, although the reliability of both is very questionable: documents can easily be planted and prisoners are not always trustworthy. When campaigning in Spain in 195 BC, Marcus Cato sent 300 men to attack an enemy post with the express aim of capturing a prisoner, who 'under torture, revealed all the secrets of his side'.[12] Frederick the Great, a true scion of the Enlightenment, eschewed torture and interrogated enemy prisoners and deserters in person, but he seldom derived anything of value from them. The peoples of most of his theatres of war – Bohemians, Moravians and Wendish Saxons – were recalcitrant and unreliable. Some prisoners are naturally loquacious, however. A Union staff officer of the American Civil War wrote years later that: 'The Confederate deserter was an institution which has received too little consideration ... He was ubiquitous, willing and altogether inscrutable. Whether he told the truth or a lie, he was always equally sure to deceive. He was sometimes a real deserter and sometimes a mock deserter. In either case he was sure to be loaded.'[13] On the other hand, Japanese prisoners captured during the Second World War, although fairly few in number, proved quite valuable sources of accurate information. Because their creed refused to accept the concept of surrender, they were never taught how to behave if they were captured.[14]

Signals intercept began the first time a messenger was waylaid, but it did not become a systematic part of the intelligence effort until technological change provided greater opportunities. When the largely forgotten hero of the Royal Navy, Thomas Cochrane, was involved in raiding the French coast between Perpignan and Marseille in 1808, one of his targets was a semaphore station. The French Garde Nationale, terrified by the approach of the man Napoleon christened *le loup des mers* ('the wolf of the seas'), retreated before the British raiding party and watched while it burnt everything. When they returned to assess the damage, they were relieved to find the half-burnt remains of their signal code books and believed the brutish British had failed to realize their value. In reality, the charred books had been planted to reassure them of precisely this, for Cochrane had in fact noted the secret wigwag code and passed it on to his superior, Admiral Lord Cuthbert Collingwood. From then on any British ships within visual range could read French signal station messages.

The invention of the telegraph opened a new dimension in communications. The first attempt at line signalling was made in 1839, but there is no record of anyone interfering with British communications during the Crimean War fifteen years later. By 1850 there were over fifty commercial telegraph companies in the United States, and during the American Civil War President Abraham Lincoln received the majority of his situation reports by this means. The first cavalry raider of that war to cut a telegraph line could be said to be the father of electronic warfare, although perhaps the laurels for inventing this new means of warfare should really go to the Confederate cavalry general John Hunt Morgan, who employed a telegraphist to intercept messages from the Union authorities and to send false ones.

On 4 July 1862 Morgan set out from Knoxville, Kentucky, on a sweep through Union-controlled Tennessee, during which he captured seventeen towns, captured and paroled 1,200 Union regulars and 1,500 home guarders, and even recruited 300 additional volunteers. Soon afterwards he broke up the Union command sent in pursuit of him and captured its commander and staff. During this time the telegraphist would sometimes chat waggishly to enemy operators, and even went so far as to complain indignantly to Washington in Morgan's name about the poor quality of the mules that were being captured.[15] Not that it was always necessary to tap the wire. J. O. Kerbey, a Union spy, would lean against the wall of a building in Richmond near the window of a Confederate signaller whose messages he could overhear being transmitted uncoded. Kerbey listened to the tap of the hammer on the transmitter and sent what he heard by a secret courier service to Washington.[16]

Guglielmo Marconi's invention of radio in the form of wireless telegraphy was soon given a military application.

The first signals were transmitted across the Atlantic in 1901 and by the beginning of the Russo-Japanese War of 1904-5 most ships in the fleets of Russia and Japan were fitted with it. The war began with a Japanese surprise attack on Port Arthur, but during the frequent repeat attacks Russian radio operators started to notice a great increase in Japanese signals in their headphones long before any sighting was made of the enemy. Thus the Russians were given warning of impending attacks and were able to put their own ships and coastal batteries on alert. When several Russian ships were dispatched from Vladivostok to launch a surprise attack on the Japanese naval base of Gensan, they intercepted radio communications indicating that Japanese ships were also heading for Gensan, and promptly abandoned their plans, which might otherwise have ended in disaster. On 8 March 1904 the Japanese attempted to carry out an attack on the inner roads of Port Arthur, planning to direct the fire of two cruisers from over the horizon by radio from a small destroyer near the coast. When a Russian wireless operator heard the exchange of signals, although he did not really understand what was going on, he instinctively pressed his transmission key in the hope of somehow interfering with them. The Japanese ships, unable to fire accurately as a result of this first example of jamming, were forced to withdraw.

The failure of the Russian admiral Zinoviy Petrovich Rozhestvenskiy to appreciate the full significance of radio communications led to disastrous and humiliating defeat at the Battle of Tsushima, but this was a sign that electronic warfare had come of age.[17] On land, radio intercept was first used effectively on 19 August 1914, when a British Army radio van at Le Cateau intercepted German messages which it passed on to GHQ. On the Eastern Front soon afterwards the German generals Erich Ludendorff and Paul von Hindenburg were able to learn of Russian troop movements by intercepting their primitive radio transmissions and consequently to destroy the Russian Second Army at the Battle of Tannenberg. Max Hoffmann later recorded that 'we had an ally, we knew all the enemy's plans'.[18]

The basic principles of intercept, direction finding and analysis were soon established, but the continuing primitive nature of the technology meant that radio was seldom employed below brigade level, where the field telephone was the main means of communication. It was not until 1915 that the British general staff, concerned at the apparent ease with which the Germans anticipated their tactical moves, realized that this too could be tapped. The Germans had developed a sensitive detector and amplifier using vacuum tubes which picked up the feeble earth currents. This led to the development of a noise jammer and in due course the British also developed their own highly sensitive amplifier, capable of detecting telephone signals up to five kilometres away. Eventually, other devices raised the level of security. The Fullerphone, for example, was practically undetectable unless the interceptor physically tapped the wire. Further advances were also made in radio direction finding, which in the 1930s was refined and developed in Britain by Sir Robert Watson-Watt to produce the first operational radar, which played a crucial role during the Battle of Britain. By 1939 the Germans had also produced an operational radar system and at this point there was a divergence between air and naval electronic warfare on the one hand, increasingly concerned with the protection or destruction of platforms (ships and aircraft), and land warfare on the other.[19]

The plethora of electronic warfare terms and acronyms can be misleading. Electronic warfare (EW) is divided into three branches: electronic counter-measures (ECM), electronic support measures (ESM) and electronic protection measures (EPM, formerly known by the unwieldy term of 'electronic counter countermeasures' or ECCM). EPM are defensive and include radio silence, code and technical measures, all designed to provide security, protect one's communications and deny the enemy information from ESM. This is electronic reconnaissance (listening), from which intelligence is derived. Once analysed and collated, this becomes signals intelligence (SIGINT, a phrase usually applied to non-battlefield transmissions such as diplomatic and other government signals), which is in turn divided between intelligence from communications systems (COMINT, or communications intelligence) and non-communications electronic systems such as radar, telemetry and guidance systems (ELINT, or electronic intelligence). ESM or electronic reconnaissance begins with searching the frequency spectrum for enemy transmissions. Once found, they can be intercepted and listened to, although they are likely to be encoded and it may not be possible to read them. Nevertheless, traffic analysis can reveal considerable information, and if they can be read they may prove invaluable. The final stage of the process is direction finding (DF). On most nets the control station will probably be the most frequent transmitter and this, combined with other information, may indicate a headquarters. ECM are designed to disrupt and attack enemy transmissions through jamming, neutralization and the feeding of false information through electronic deception (ED).

The importance of radio in modern war means that the deceiver seeks to dominate the enemy's use of the electromagnetic spectrum, so that false information can be conveyed and genuine information denied. It is also imperative to control use of the electromagnetic spectrum by friendly forces. British commanders in the Middle East during the Second World War became paranoid about spies in and around GHQ in Cairo, who it was believed were leaking tactical plans to the Germans. It was the Germans' use of radio intercept that enabled them to divine British moves, a task made considerably easier by the laughable naïvety of British operators who used 'veiled' speech rather than proper voice procedure. They believed, for example, that references to cricket and hunting (for example, 'returning to the pavilion for tea' as a euphemism for replenishment of fuel and ammunition) were sufficient to confuse the listening Germans.[20] Only when 9th Australian Division overran the German intercept unit at Tel el Eisa in July 1942 did the extent of intelligence that the Germans derived from this source become apparent. However, the consequences for deception of this rather distasteful discovery were considerable. The

Lowe dutifully reported all he saw to McClellan's head-quarters using a telegraph carried in the basket (and greatly assisting 'Prince John' Magruder in his peninsular deception). But the potential of the aeroplane when it arrived failed to convince everyone. The eminent French general Ferdinand Foch declared in 1910 that for army use 'l'avion c'est zéro!' During the following year's manœuvres, however, his colleague Joseph-Simon Galliéni captured a colonel of the Supreme War Council and his entire staff thanks to a reconnaissance aeroplane.[23] The same year the Italians made the first use of powered aircraft in war against the Turks in Libya. Capitano Carlo Piazza borrowed a camera from the photo section of the Engineer Corps on 23 February 1912, and the results were so impressive that his colleague Ricardo Moizo immediately followed suit. While they produced few prints, they did highlight inaccuracies on maps and the possibilities for the future were demonstrated.[24]

The British Expeditionary Force that went to France in August 1914 was accompanied by four squadrons from the Royal Flying Corps, whose sole purpose at the time was reconnaissance. Tactically, this meant artillery observation and the location of enemy batteries, reporting trench locations and in due course hampering the enemy's attempts to do the same. Strategic reconnaissance in 1914 meant anything beyond five miles of the front, and the limitations of simple observation soon became apparent. The ability of the camera to record information accurately and reliably was soon put to use therefore, initially by No. 3 Sqn, which had pioneered photographic techniques before the war. Lieutenant G. F. Petyman took the first five exposures over the German lines on 15 September. By the following year the lavish equipment and centralized facilities available to the French compared most unfavourably with the *ad hoc* arrangements made by the British. Major W. G. H. Salmond, officer commanding No. 3 Sqn, recommended that a similar organization be adopted by the British and an experimental section was formed. In due course a magazine was developed that enabled exposures to be made in rapid succession and stereoscopy greatly enhanced the value of the resulting photos.[25] (If approximately sixty per cent of overlap is achieved on two prints, a stereoscope will permit three-dimensional viewing, from which far more information can be derived.) However, it was a long time before the techniques of photo reading and interpretation were fully explored and appreciated; in the meantime the RFC had to cope with anti-aircraft fire and the scourge of the Fokker fighter. Nevertheless, by 1918 every major application of photographic reconnaissance to be used for the next fifty years had been tried and tested.

During the Second World War Britain's Photographic Reconnaissance Unit (PRU) divided interpretation into three phases. First phase meant immediate reporting of new items such as ship and aircraft movement, rail and canal traffic and bomb damage assessment. Second phase reports were produced within twenty-four hours and covered general activity, and were collated with the day's accumulated coverage. Third phase was the very detailed statements prepared for specialist requirements, usually on fixed installations such as airfields, factories and important experimental facilities. As the war progressed, third phase was dealt with by the Central Interpretation Unit (CIU) at Medmenham, near Henley, which became expert in divining the strategic implications of what it saw, enabling the discovery and subsequent bombing of targets such as the V-weapon test site at Peenemünde. Photography also allowed the state of construction of U-boats at Kiel and Bremen to be measured. David Brachi, one of the RAF's photo specialists, remarked at the time that 'the Germans are so methodical about their camouflage that once you get to know their methods you can tell quite a lot from the camouflage itself'.[26]

By comparison, while in 1939 the Germans possessed far more photo interpreters (PIs) than the British, they have often been criticized for not using stereoscopes in their day-to-day work and for relying largely on non-specialist NCOs, a reflection of their view that photo interpretation was a mechanical process.[27] But there was a significant difference in approach. While the RAF concentrated on strategic targets, with the tasking coming from a high level such as Coastal or Bomber Command, the Luftwaffe was geared until as late as 1943 towards the tactical demands of *Blitzkrieg* and towards supporting the army in the fluid and fast-changing environment of a battlefield. The RAF's PIs thus became experts in strategic subjects such as shipping and airfields, while the Luftwaffe concentrated on battlefield terrain and fortifications. (The RAF also provided tactical reconnaissance, or 'Tac R', in support of the army, mainly from specifically tasked army

JON LATIMER

co-operation squadrons.) This lack of a strategic dimension to Luftwaffe operations proved a serious drawback as the war progressed, and contributed greatly to Germany's ultimate defeat.[28]

Security

Security is as fundamental a principle of war as intelligence. Frederick the Great once declared that if he thought his coat knew his plans, he would take it off and burn it. Detailed knowledge of the enemy's reconnaissance and intelligence capabilities are vital if one's secrets are to be preserved.

Field or operational security involves the concealment of one's own strengths and intentions from the enemy. Thus Napoleon's cavalry, while gathering information on the enemy, also prevented the enemy from reciprocating: acting as a moving screen, it disguised Napoleon's operations from enemy patrols and protected his lines of communication and operational base where the depots, hospitals and parks were situated. Before a campaign opened, Napoleon habitually lowered the curtain of military security. The press, so often a source of information about impending military moves in the eighteenth century, was ruthlessly controlled and 'tuned' to produce the information that Napoleon wished the enemy to have. Weeks before any move the frontiers of France would be closed to foreigners and the secret police would redouble their activities in watching suspects. At the same time elaborate deception schemes and secondary offensives would

64

be devised and implemented to confuse the foe and place him off balance. Thus Napoleon employed methods that were to become common in twentieth-century warfare.

For the purposes of both security and deception Napoleon was in the habit of continually altering the composition of his major formations – adding a division here, taking away a brigade there, creating an occasional provisional *corps d'armée* for a special mission in mid-campaign – measures that served to confuse the enemy still further. On 16 October 1805, for instance, Austrian intelligence learned that outside Ulm Maréchal Jean Lannes's V Corps comprised the infantry divisions of generals Oudinot and Gazan and the light cavalry of Treilhard. But from the 24th of the same month Lannes's command included two more infantry divisions transferred from Ney's and Marmont's corps and no fewer than three more cavalry formations from Murat's cavalry reserve. No sooner was this intelligence discovered and digested by the enemy, however, than it was completely out of date, for the moment the French advance passed the River Enns, the same administrative and operational flexibility enabled Napoleon to withdraw three infantry divisions from Lannes and form them into a new provisional corps (the VIII, under Mortier). Thus at no time could the enemy rely on the accuracy of information concerning the strength of the French or the placing of their units.* As the distance between the two

---

* Ironically, the French were similarly deceived in 1914 by the German habit of giving reserve formations the same numeral as their parents. So, for example, an army including the IV Corps would be grouped together with IV Reserve Corps.

sides closed, security became more difficult to maintain and both would receive a stream of information – some of it misleading, to be sure, but most of it relevant. Then, when the 'veil was torn', Napoleon would rely on speed of movement, extending the length of marches and forbidding all foraging, and then the jealously conserved supplies of the ration convoys would be distributed.[29]

The larger a proposed operation, the more difficult that concealment becomes. Operational security is most effective when applied systematically; it must be directed from the highest level and must concentrate on critical activities, identifying what indicators an enemy will look for and what information these might convey to the enemy (bridging equipment, for example, will obviously suggest an intention to cross a river). It must also take account of the enemy's reconnaissance, surveillance and target acquisition capabilities, so that measures can be designed to neutralize these. (There is no form of camouflage more effective than putting out the enemy's eyes.[30]) Comprehensiveness and timeliness are equally important here: assessments must be made before and during an operation and continuously revised, since any protection measures taken must appear a normal part of activity: routines can thus both aid security and provide a basis for deception. Finally, as in every military activity, the plan must be capable of change at short notice.

The very identity of a general must be subject to security, and deception can aid this. Hannibal was well aware of the fickleness of his Celtic allies and, having only recently established friendly relations with them, he was on his guard against attempts on his life. He therefore had a number of wigs made, and these he constantly changed, along with his style of dress, so that even those who knew him well had difficulty recognizing him.[31] Similarly, a general's personal routine can be an indicator of forthcoming operations. General Sir Archibald Wavell and his field commander, Lieutenant-General Richard O'Connor, took enormous care to ensure that security was watertight for their great offensive Operation COMPASS in western Egypt in December 1940. Only those who absolutely needed to know were involved in the planning, and when a rehearsal was necessary, nobody taking part knew the real purpose. A second training exercise was then scheduled and just forty-eight hours before this was due to start, operational orders were issued instead of training instructions. Wavell himself conspicuously attended the races in Cairo with his family on 7 December, and then attended a dinner party for senior officers that evening. The operation started the following night. The Egyptian prime minister, Hussein Sirry Pasha, who took great pride in 'having sources who keep me informed of all that goes on', congratulated Wavell 'on being the first to keep a secret in Cairo'.[32]

COUNTER-SURVEILLANCE

Before offensive deception measures can be planned, friendly surveillance effort must be directed towards establishing the type and density of the enemy's sources and towards looking for weak spots. As an aid to security and an integral part of the information battle, counter-surveillance, involving all those active and pas-sive measures taken to prevent hostile surveillance of a force or area, forms the first category or level of decep-tion. These essentially defensive measures are not, how-ever, strictly deception techniques in their own right. Deception aims to mislead the enemy into adopting a predictable course of action that can subsequently be exploited. Lack of information and confusion are natural states on the battlefield, and reinforcing these conditions for the enemy by counter-surveillance can contribute to surprise, security and deception. But an enemy deprived of all intelligence or faced with ambiguous information may react unpredictably, and his actions may not nec-essarily be exploitable. Nevertheless, denial of genuine information is always an important objective and confu-sion may in some cases be a useful method of supporting deception by undermining the enemy's intelligence effort.

Active counter-surveillance measures include attack-ing enemy reconnaissance forces and passive ones include camouflage, the use of smoke, absence of movement, radio silence and all the other measures taken to conceal the presence of forces or installations such as supply dumps. In the face of modern high-technology surveillance equipment such as radar, thermal pointers, night-viewing devices and drones (remote piloted vehicles carrying cameras and other devices) this concealment is extremely difficult, but what cannot be hidden or disguised can be misrepresented. The priorities for defensive deception measures should be related to the enemy's reconnaissance priorities and capabilities, underlining again the need to understand as far as possible the enemy's intelligence cycle.

Camouflage is a key element of counter-surveillance. In March 1918 the British General Staff issued a pamphlet called *The Principles and Practice of Camouflage*, which distilled four years experience of modern warfare. It stated quite clearly that *'Deception,* not *concealment,* is the object of camouflage.'[33] It defined camouflage as 'concealment of the act or fact that something is being concealed' and continued, *'deception* is the essence of it.' The word is derived from the French slang word *camoufler* ('to disguise') and was first used by hunters. There are isolated instances of its use in ancient and medieval warfare, but for most of history warfare was largely confined to close quarters and it was not until the advent of the rifle as an effective military weapon in the eighteenth century that camouflage began to be developed, initially by irregular units fighting in North America. The first unit to be uniformed entirely in green was the New York Militia, in 1795. In 1797 a 5th Battalion was raised mainly from Germans for the 60th (Royal American) Regiment, and became the first British unit to wear green.

Certainly a green-clad soldier would be less conspicuous a target than one in scarlet (although the British soldier's red coat was less conspicuous than it might seem, since weathering soon reduced it to a shade of brown). But the effective range of the musket was only around 100 yards and even the rifles of the period were only effective to around 300 yards. Moreover, since the muzzle-loading technology meant that the rate of fire of the musket was seldom more than three rounds per minute, even in the hands of well-trained troops, in order to generate effective firepower it remained necessary to manœuvre in close order and fire in volleys. In any case, experiments carried out by Captain Charles Hamilton Smith early in the nineteenth century, involving rifle shots at a range of 150 yards, proved that the least conspicuous colour was actually the light grey uniform worn by Austrian *jägers*. Green was actually chosen because of its associations with the role of hunter played by those units equipped with rifles. The formation of a regiment of riflemen, the famous 95th, saw their dress being of the same 'rifle green' as the 5th Battalion, 60th Regiment.[34]

Following the Napoleonic Wars, part of the British Army was almost continuously engaged in India. From the 1830s onwards there were also a number of small wars in southern Africa, where the troops drew on the experiences of European settlers and were quicker to adapt to bush warfare. Many officers wore hardly any uniform at all, and dressed for the bush from stores in frontier towns. In 1851 the 74th Highlanders discarded their red coatees in favour of brownish-grey canvas smocks, albeit for reasons of serviceability rather than camouflage. The Corps of Guides were raised from among the Sikhs by Harry Lumsden in 1846, following the Sikh Wars. In 1848 they were dressed in khaki (from the Hindustani *khak,* meaning 'dirt') or 'drab' as it was officially called, introduced by Lumsden and William Hodson. During the Indian Mutiny of 1857-8 the first British regiment to adopt khaki was the 52nd Light Infantry, whose normal white summer clothing was dyed in the local bazaar before the regiment left for the siege of Delhi. The 61st Regiment dyed their kit a sort of bluish- brown at about the same time and by the end of the mutiny most regiments had followed the example with whatever came to hand, including earth, tea and curry powder.[35] On 21 May 1858 the adjutant-general announced 'that for the future, the summer clothing of the European soldiers shall consist of two suits of "khakee"'.

In 1868 an expeditionary force sent to free European hostages being held by the mad emperor Theodore of Ethiopia saw the first use of khaki outside India, and during the Second Afghan War (1878-80) white coats were again stained with tea. Khaki drill service dress was formally introduced into the Indian Army in 1885, and in 1896 a standard brown khaki was introduced for all foreign service outside Europe. Soon afterwards, as the experience of the Second Anglo-Boer War reinforced the need for camouflage in the face of the awesome power of the modern rifle and smokeless propellant (on many occasions British troops had been pinned down by invisible enemies firing from up to a mile away), scarlet was banished to ceremonial duties for ever more.[36]

The US Army, which had similar experiences during the Spanish-American War (1898-1901), also introduced khaki for all occasions other than ceremonial and most of Europe soon caught up; prompted by their newly acquired possessions overseas and the experiences of the police and *schütztruppen* raised to guard them, the Prussians in 1908 adopted *feldgrau* ('field grey'); the rest of Germany followed suit in 1910 and the colour became their hallmark between 1914 and 1945. The Italians chose a grey- green in 1906 and the Russians, also as a result of experience during the Russo-Japanese War, adopted khaki in 1908.[37] The notable exception was the French. In 1912 the French Minister for War, Adolphe Messimy, visited the Balkans, where he was impressed by the way the dull-coloured uniforms in use there blended into the landscape. He returned to Paris and proposed a similar transition for the French Army, which still wore basically the same uniform as it had in the 1830s. The reaction was one of total indignation that anyone should so much as dare to tamper with the glorious traditions of the French Army. In government hearings the offensive spirit engendered by the traditional blue tunics, red kepis and red trousers was deemed indispensable. As one former war minister declared: 'Les pantalons rouges, c'est la France.' They were retained, and Messimy later noted that this 'blind and imbecile attachment to the most visible of all colours was to have cruel consequences'.[38] When war came soon afterwards, the French Third and Fourth Armies ploughed headlong towards Germany into the teeth of withering fire. Thousands of Frenchmen paid the price of the lesson of camouflage, which would come to full maturity during the First World War.

The Waffen-SS were the first to develop clothing with disruptive patterns and the Germans, and to a lesser extent the Soviets, made extensive use of this sort of material during the Second World War. The Western Allies made only limited use of it (notably British parachutists and US Marines), but practically every army in the world has subsequently adopted it in one form or another.[39] However, patterned camouflage of equipment to prevent easy observation, especially from the air, was a much earlier innovation. During the First World War patches of black were added to mottled greens and browns when it was found this helped to break up the shape of a gun or vehicle. In France the idea was that of a fashionable Parisian portraitist serving in the artillery, Guirand de Scevola. In 1914 he painted some canvas sheets to throw over guns when they were out of action. Pablo Picasso, travelling through Paris and seeing a camouflaged gun declared: 'It is we that have created that.'[40] The GQG (French High Command) were so impressed by this idea that they gave de Scevola a commission (in both senses) and recruited other painters, including André Segonzac and Jacques Villon, to form a mobile corps whose task was to travel the line camouflaging artillery, airfields and observation posts. By 1918 some 1,200 men and 8,000 women were employed in workshops under de Scevola's supervision; the artist himself was always elegantly dressed in white gloves.

Prompted to a large extent by another artist, Solomon J. Solomon, the British soon adopted similar measures. One of Solomon's first tasks was to create observation posts that looked like trees. Responsibility for camouflage was given to the Royal Engineers, which formed a Special Works Park, headed in 1916 by Lieutenant-Colonel Francis Wyatt MC; the painters involved included Henry Paget, Walter Russell and Alan Beeton. By the following year the park had a strength of 60 officers and 400 other ranks, employing hundreds of French women to garnish camouflage nets and supplying the needs of four

British armies in the field. The Germans employed the avant-garde artist Franz Marc, who, in a break from front-line service with the cavalry, was employed painting what he called nine 'Kandinskys' on military tarpaulins. Marc himself subsequently returned to the front line and was killed at Verdun in 1916.

The arrival of peace in 1918 put camouflage to the back of British military priorities during the 1920s and 1930s. However, following a report by Brigadier Andrew Thorne, commanding 1st Guards Brigade, the War Office commissioned Frederick Beddington to investigate dis-ruptive paint patterns for vehicles and when the Second World War broke out, Beddington was put in charge of the Camouflage Experimental Section. A camouflage fac-tory was set up at Rouen, but everything was abandoned following the German breakthrough in May 1940. With the Army's deficiencies so brutally exposed and camou-flage and deception suddenly vital, the section was turned into the Camouflage Development and Training Centre (CDTC) at Farnham. Painters such as Edward Seago, Frederick Gore and Julian Trevelyan, along with designers and architects, were turned into staff officers (camouflage) and posted to headquarters throughout the Mediterranean and Far East theatres. The Americans by contrast orga-nized camouflage battalions as combat units rather than merely producing specialist staff officers, and attached one such unit to each army. The driving force behind them was Lieutenant- Colonel Homer Saint-Gaudens, who had played an important role in instilling camouflage disci-pline into the 'doughboys' in the First World War.[41]

Other products of the CDTC included the West End magician Jasper Maskelyne and the film-maker Geoffrey Barkas, both of whom wrote accounts of their exploits. Barkas was posted to GHQ Middle East, where he became Director of Camouflage. There he rapidly decided that camouflage was not something that could be confined to specialists; all ranks and all arms would require training in the technical aspects, and senior officers would need to understand what was meant by disruption and counter-shading, and require training in the interpretation of air photos and deception. Furthermore, large-scale workshop facilities were needed to produce the vast quantity of necessary materials. A Middle East version of the CDTC was set up at Helwan, near Cairo, and No. 85 (South African) Camouflage Company provided the training centre and six mobile detachments. Together with No. 1 Camouflage Company, Royal Engineers (a large unit of 7 officers and 267 men, formed from British and Palestinian Jews), they laid the basis for camouflage to be employed not merely as a passive, defensive measure, but for active deception. Many lessons that the British had learned some twenty-five years' before when the Egyptian Expeditionary Force fought the Turks in Palestine, had now to be relearned. However, the Middle East would provide the British with a proving ground for the development of tactical and subsequently operational deception techniques that would be employed to an unprecedented degree and considerable effect during the second half of the war, and which would in due course enable the implementation of effective strategic deception.

CHAPTER THREE

THE PRINCIPLES OF DECEPTION

'There is no more precious asset for a general than a knowledge of his opponent's guiding principles and character, and anyone who thinks the opposite is at once blind and foolish ... In the same way the commander must train his eye upon the weak spots in his opponent's defence, not in his body but in his mind.'

Polybius

Focus

DECEPTION MUST ALWAYS be aimed clearly at the mind of the enemy commander, at the man who makes the decisions, whether it be the head of state of a country or an ordinary soldier. All human beings are prone to certain psychological vulnerabilities. Our learning processes are conditioned by our physical, cultural and social environment, and we tend to compare whatever situation confronts us with the templates, formed by experience, through which we view the world. This is particularly true of military organizations, where rank and experience count for more than practically anything else. In *The World Crisis* Churchill comments that

> the firmly inculcated doctrine that an Admiral's opinion was more likely to be right than a Captain's and a Captain's than a Commander's did not hold good when questions entirely novel in character, requiring keen and bold minds unhampered by long routine, were under debate.

It is common practice to tell leaders what it is believed they want to hear. Churchill also noted that 'the temptation to tell a chief in a great position the things he most likes to hear' is the commonest explanation of mistaken action. This tendency is noticeably stronger among totalitarian regimes, as Hitler's and Stalin's sycophantic adherents demonstrated to their cost. It is not surprising therefore that commanders are sometimes led to jump to conclusions, either prematurely or against the run of evidence. The mind is susceptible to being lured towards particular information, and misled by its own preconceptions. In this context the intelligence chief may be an important conduit by which the deception is conveyed, but ultimately the target must be the enemy commander.

ACTION

In 1940 General Sir Archibald Wavell, Commander-in-Chief Middle East, formed a specialist unit under

Lieutenant-Colonel Dudley Clarke which became known as 'A' Force. Its purpose was to devise and conduct deception operations, and its first operation was code-named CAMILLA.[1] Wavell wanted the Italians in Abyssinia to think he was about to attack them strongly from Kenya in the south, driving on into occupied British Somaliland, from where operations would be conducted into Abyssinia itself. This was in order to draw the Italians away from the north, where his genuine main effort was to be made from Sudan. Plentiful resources were made available to 'A' Force, but the deception went, if anything, too well. The Italians retired, presumably because they believed the notional attack from the south was likely to succeed and that by withdrawing to a shorter line they could create a stronger defence. At the same time they sent reinforcements to the northern flank, where the 4th and 5th Indian Divisions eventually had a fierce fight to overcome the formidable defences at Keren.[2]

Some time in early 1942 Clarke, now promoted colonel, sat down to write a summary of the lessons his unit had learned. 'It is important to appreciate from the start' he wrote,

> that the only purpose of Deception is to make one's opponent ACT in a manner calculated to assist one's own plans and to prejudice the success of his. In other words, to make him *do* something. Too often in the past we have set out to make him THINK something, without realizing that this was no more than a means to an end. Fundamentally

it does not matter in the least what the enemy thinks: it is only what line of action he adopts as a consequence of his line of thought that will affect the battle. As a result we resolved the principle that a commander should tell his Deception staff what he wants the enemy to DO ... while it is the duty of the latter to decide, in consultation with the Intelligence Staff, what he should be made to THINK in order to induce him to adopt the required course of action.[3]

The deceiver's principal aim is thus to support the commander's mission and his concept of operations. A secondary goal might be to degrade the enemy's reconnaissance and intelligence capacity, but that is of relatively minor importance to a deceiver: too much ambiguity can mask the story.

Deception is created by manipulating perceptions. The first task is to identify a bias or, if necessary, to create one. It is always much easier to reinforce a perception than to change one, and if one knows what an enemy expects will happen (or better still, hopes for) then any deception will be on firm foundations. Then one feeds information, true, false, partially true and misleading, in order to reinforce that perception with the aim of inducing a reaction. British doctrine maintains that deception has four main objectives. The first is to provide a commander with freedom of action to carry out his mission, by deluding the enemy as to his intentions and by diverting the enemy's attention away from the action being taken, in order to achieve the aim. The second is to mislead the enemy and persuade him to adopt a course of action that is to his disadvantage and can be exploited. The third is to gain surprise, and the fourth is to save the lives of one's own troops.[4] The first of these objectives corresponds broadly with counter-surveillance, the largely defensive measures discussed above, such as camouflage. Since the Second World War there has been a growing tendency to regard these as the be-all and end-all of deception, but the active, offensive measures to which Clarke refers require more than just a lick of paint.

Co-Ordination and Centralized Control

Creativity and originality are absolutely fundamental in planning any deception, but thereafter the deceiver should be guided by principles regarded by Clarke as no more than common sense. It is incorrect to think that deception is a function of the intelligence branch of the staff. Control should lie instead with the operations staff. 'Op[eration]s are the user and dictate the Object, direct the tempo of the plan and decide when it must be replaced.'[5] Although intelligence has a crucial role to play in preparing and monitoring the effectiveness of any deception, it must always be controlled by the operations branch since they have the executive power to implement it. In order to serve the commander's purpose, deceivers must be in constant touch with his thoughts. This is a two-way process, and while the deception staff works within the operations branch its head should have direct access to the commander.[6]

A successful deception will occur only if all the staff branches responsible for their various aspects of operational planning are properly co-ordinated. The overall planning must be an operations matter, but the latter will have to work very closely indeed with the intelligence branch and possibly with logistics and artillery, and particularly with engineer and communications staff. Command must be exercised at the highest level but – unlike, for example, the operation of artillery, where command is exercised at the highest level but control is devolved downwards to achieve flexibility – it is imperative that deception be not only centrally commanded but also centrally controlled. Modern professional armies like to promote initiative in junior ranks, and when fighting a battle this makes eminent sense. But when painting a complex and delicate picture, it is vital that direction is closely adhered to.

During the First World War the Admiralty's Director of Naval Intelligence was Admiral Sir Reginald Hall, known as 'Blinker' Hall because of a slight tick in one eye. He was something of a maverick and fascinated by anything to do with spies, deception and what might be called 'dirty tricks'. This aspect of the subject is far removed from the genuine, methodical and painstaking reality of most intelligence work, but the two facets are often inextricably entwined.[7] Hall is best known for the brilliant signals intelligence work conducted in Room 40 by his cryptographers, but he took this further by having printed a 'Secret Emergency War Code' book, which he then allowed to fall into German hands by means of a diplomatic courier in neutral Rotterdam. He then encouraged the Germans to have confidence in the code by using it for seemingly important signals. It was then used to deceive them, first over a plan to 'invade' Sylt, an island in the north Friesian chain off the coast of Germany. Second, he put out a story in August 1916, when the Battle of the Somme was raging, that the British intended to invade northern Belgium with the aim of persuading the Germans to draw troops away from the battle front to cover this threat. With great skill Hall allowed the Germans to piece the story together themselves and used the secret code to instruct warships due to escort the 'invasion' to form up in three groups based on Dover, the Thames estuary and Harwich. As these were the main supply ports for the troops in France, there was no shortage of shipping for a Zeppelin captain or a pilot to report, and since the only recipient of these signals was the German Y-Dienst radio intercept service, it did not interfere with actual naval operations. Hall added to this deception with a special printing of the *Daily Mail,* a few copies of which were sent to the Netherlands. This was soon followed by another edition from which a prominent article was removed, as if it had been censored, suggesting preparations along the east coast of England involving flat-bottomed boats. The Germans responded by deploying troops to the threatened area.

The problem was that no one in the Admiralty had informed the War Office of this scheme, and the German troop movements towards the coast of Belgium led to the worst invasion scare of the war. Without a co-ordinated and reliable intelligence and counter-intelligence network with which to monitor the German reaction, Hall could not be certain that this was only a response to his deception, and he therefore kept quiet about his ruse. While it is easy to see the faults in such a scheme, it should be remembered that co-ordinated strategic deception was in its infancy and that Hall's primary concern was naval. There was no Joint Planning Staff to advise him and as DNI he was a long way from the centre of operational planning. But both of these lessons were learned in time for the real invasion of Europe in 1944.[8]

PREPARATION AND TIMING

Preparation is crucial. A poorly planned deception may be worse than no deception at all. And in this process timing is possibly the most critical factor.

> Every Deception Plan must be given time to work. It is no good telling a Deception Staff to try and influence an enemy 'at once'. The Plan must be aimed at making him act in a favourable manner only at some selected future date, when its implementation has had a fair chance of exerting some effect.[9]

There needs to be enough time to develop the concepts and to ensure the deception planners and implementers have time to paint the picture. Planners must be aware of the time a given measure will take to produce the desired effect in the mind of the target, and for the target to react as desired.

The planning process must follow a logical progression. First, the commander must decide what he is really going to do. Deception only becomes possible when operational intentions have been determined.[10] Second, a cover plan should be created based on the principles of credibility and timing. This must follow the real plan, not dictate it: any attempt to fit the reality to the deception is doomed to failure.[11] Especially at the higher levels (operational and strategic), deception achieves results by a steady increase of momentum and it must be appreciated that it takes time both to gain momentum and to lose it. Employing deception at the eleventh hour may not only be too late to succeed but may actually interfere with the genuine plan.[12] The reaction time of the target's sources must be calculated and the timing of the deception planned accordingly. Timing should be logical so that combat indicators follow the sequence the enemy would expect, and the deceivers need therefore to understand the enemy's intelligence and decision-making process. Ideally they will also have some sort of feedback on the progress of their effort. This further reinforces the need for a single staff element to be responsible directly to the commander for proper coordination of those implementing and evaluating the deception. This staff officer ensures that the real and cover plans are complementary and mutually supporting. As Dudley Clarke learned through difficult experience, 'deception will pay its best dividends when both planning and implementation by *all* methods is made the responsibility of one controlling mind.'[13]

Security

A deception planner who inserts his message into too *[handwritten: Too much opsec is probably bad.]* many channels risks misleading his own side and alerting enemy analysts. In order to ensure that his message will be received, he must begin by making sure there is a low number of channels that the enemy will find productive or promising; in other words, good deception begins with good security.[14] Two levels of security are required in any deception plan: first, the genuine operation plan must be secure so that the enemy cannot determine one's true intentions; second, the deception plan must itself be equally secure, if not more so. The mere existence of a deception plan, let alone the details, should be known only by those who need to know. In the early period of the Great Patriotic War (as the Soviets called the Eastern Front of the Second World War) loss or careless transmission of planning documents compromised Soviet operations in general and *maskirovka* ('deception') in particular. They therefore implemented fierce security restrictions on the numbers of planners and documents involved in any operation, and communicated only what a subordinate needed to know, and only when he needed to know it (and never why he was engaged on any particular task, however odd it might seem).[15] Dudley Clarke always played his cards very close to his chest but his manner was so pleasant he could get away with it. If anybody showed interest in his work he would start telling funny stories on a completely different topic.[16]

Breaches of security, however, need not compromise either an operational or a deception plan. Some leaks may not be noticed by a target, either because his intelligence fails to pick them up or because his preconceptions may induce him to misinterpret them. It could even be argued that the bigger the leak, the less likely the target is to believe it: he may suspect that the leak itself is a deception.[17] When a German military plane carrying Major Helmut Reinberger of the Luftwaffe and containing plans for the intended German invasion of the West, including Belgium and the Netherlands, made a forced landing at Mechelen-sur-Meuse in Belgium on 10 January 1940, the reaction among both British and French High Commands was that the documents were a plant.[18] A deceiver should therefore avoid such 'windfall' inputs unless they are very cleverly disguised and part of a wider plan. If, on the other hand, a genuine deception plan is discovered (and it is hardly something a commander would want to risk), security measures should ensure that even in this case the enemy does not discover the commander's real intentions, preferably by leaving several interpretations open.

CREDIBILITY AND CONFIRMATION

A deception will not succeed in its aim if the enemy does not believe either the source or the cover plan. In the first instance, for example, if deceivers rely on double agents or false radio traffic, then the double agent must appear to the enemy to be reliable and false radio traffic must conform to normal patterns. If the enemy has any doubts as to the reliability of his sources, for whatever reasons, the deception is less likely to succeed. And for the cover plan to be credible the deceiver must be capable, in the target's eyes, of doing what the lie suggests he will do. What is actually possible is less important than what the enemy believes to be possible.[19] During 1943 the Allies tried to persuade the Germans that they intended a cross-Channel invasion of the continent in September under the code-name COCKADE, principally with the aim of drawing the Luftwaffe into battle. The German response was extremely disappointing: they made no effort to reinforce the French coast and continued to dispatch reinforcements to the Eastern Front because their intelligence reported that 'the resources in Great Britain are insufficient to permit any attempt to invade the continent this summer.'[20]

The credibility of a cover story can be enhanced when the story is confirmed by a variety of sources. Good intelligence will always seek corroboration of information and deceivers must seek to provide it. For example, if an air photograph reveals what looks like an enemy defensive position, this could be verified by a ground patrol going out and seeing men moving about on it. If subsequently the sounds of battery-charging generators and field cookers are heard, heat sources are detected with a thermal pointer (a gadget that indicates the direction from which the heat is coming), radio direction-finders confirm radio traffic emanating from it, and the smell of cooking is borne on the breeze, it might be reasonable to assume that the position is not a dummy one. However, all of these sources can be simulated and if the deception is skilfully planned and executed, it will require equal skill, allied to knowledge and experience, in order to detect it.[21]

This is particularly true if the pieces of information are allowed to reach the enemy in such a way as to convince him that he has discovered them by his own efforts or by accident. If he puts them together himself, given the nature of 'noise' in the intelligence cycle, he is far less likely to believe that the intended picture is a deception.[22] Similarly, if he has had to work hard for his material he is more likely to defend its interpretation with conviction. However, this does not mean blanket coverage is either necessary or desirable. It is both easier and more effective to use a few proven channels rather than to dissipate energy on a wide variety, the value of which may be dubious.[23] The deceiver should select those sources which are most easily fooled while neutralizing the remainder by active counter-surveillance so that the target ignores, twists or explains away any details that do not fit. The British learned this early on during the Second World War.

The sources themselves are of vital importance. Intelligence sources are usually ranked according to their reliability, and a few 'reliable' sources will probably carry more weight than many 'unreliable' ones. The Germans believed rather naïvely in the efficacy of spies in wartime, whereas the British accepted fairly early on the immense difficulty and danger inherent in espionage in enemy territory in time of war.[24] This German reliance on agents in Britain, reinforced by the fact that they had few alternatives, tended to make them overlook errors rather than question the validity of this source. Consequently, since all the agents they sent to Britain were turned against them as double agents, and with aerial reconnaissance unable to provide corroboration, it was easy for them to be deceived by the Allies.[25] (British success in turning German agents in the UK did not, however, prevent the entire British network operating in the Netherlands from being turned and run by the Germans, an episode of astonishing incompetence.)

Dudley Clarke had another principle: the lie was so precious that it should always be attended by a bodyguard of truths. By knitting the cover plan into the less critical details of the real plan wherever possible, not only could the target confirm the story for himself but the likelihood of leaks was reduced. Truths made up at least eighty per cent of the information that 'A' Force fed to the enemy, even if they stemmed from dummy fleets and tanks or false divisional signs painted conspicuously where Allied troops were practically non-existent. Thus eighty per cent of the material was confirmable by the Germans from other sources of information, and this made it easy to create a lethally misleading picture.[26] The deception for Operation TORCH (the invasion of North Africa in November 1942) was found to be effective where the build-up in Gibraltar had been passed off as aimed at the relief of Malta, while threats to France and Norway were found to be less credible, underlining the fact that 'cover stories ought to be as near the "real thing" as ... safely possible'.[27]

FLEXIBILITY

Von Moltke observed that no plan survives contact with the enemy. In war uncertainty is the only thing that can be guaranteed with any certainty, and flexibility (itself a principle of war) is of particular importance to any plan, whether operational or deceptive. Deception plans should take advantage of elements, such as terrain and weather, and as real conditions change so must the lie, if it is to avoid exposure.[28] The ability to gauge feedback from the target is invaluable in this respect, since the effectiveness of certain strands of deception, or of the developing thoughts and intentions of the target may present an unexpected opportunity, almost certainly fleeting. Given that feedback is itself subject to the same problems of evaluation as any other intelligence, it is rare to have available such a reliable source as the decrypts of the high-level German signals intercepts code-named ULTRA used by the British during the Second World War. Even then, decoding such a valuable source in time to make use of it was a knife-edge business, but it did give a priceless insight into the reaction of Hitler and his staff to British deception, and made the more complicated plans possible.[29] It also enabled the flexibility by which, when the Germans retained large forces in the Pas de Calais after the Normandy landings, the Allies were able to spin out the deception plan FORTITUDE SOUTH for almost two months.

CHAPTER FOUR

THE METHODS
OF DECEPTION

'I have always believed in doing everything possible in war to mystify and mislead one's opponent.'

A. P. Wavell

I n their book *Strategic Military Deception* D. C. Daniel and K. L. Herbig identify two types of deception: the 'ambiguity-increasing' variety (or A-type) and the 'misleading variety' (or M-type). A-type deceptions aim to hinder the identification of the true aim long enough to promote inaction in the enemy (for example, by delaying mobilization or deployment of reserves). The lies told must be sufficiently plausible and consequential to demand attention, or else force the opponent to cover multiple contingencies and spread his resources so thinly as to be vulnerable to a concentrated strike. Examples of this type are Hitler's invasion of the USSR in 1941 and Egypt's assault across the Suez Canal in 1973. The aim of an M-type deception, on the other hand, is to reduce ambiguity by suggesting that one particular option is most likely, thus inducing the target to concentrate operational resources in the wrong place, as with the Allies' threat to the Pas de Calais before the Normandy invasion of 1944. In practice, however, the distinction between these two types is often blurred.

Virtually all stratagems and manœuvres of war are variations on a few simple themes.[2] There are five principal categories of deception, which might be enlisted singly or in combination to produce either an A-type or M-type response. The first category is counter-surveillance, considered above. However, there will be times when a commander deliberately wants to attract the enemy's attention in order to mislead him. The second category of deception therefore comprises displays that are deliberately intended to catch the enemy's eye, and includes all decoys, mock-ups, dummy positions, equipment and obstacles, simulated tracks of wheeled vehicles and armour, smoke and heat sources, radio traffic and electronic emissions. These can be used to portray a unit that does not exist, to give the impression that there are powerful forces in an area where there are actually very few, or to disguise the true nature and strength of a ship or unit that cannot be concealed. The third and fourth categories involve the manœuvring of forces. The third category, feint operations, comprises movements made with the object of deceiving the enemy as to the timing, weight or direction of the main attack. Diversionary raids have the same effect, as will a feigned withdrawal. In the fourth category are demonstrations, which are similar to feints but with the essential difference that a demonstration is a show of force on a front where the deceiver has no intention of fighting. The fifth category comprises ruses: tricks, strategems or cunning stunts designed to deceive the enemy.

DECEIVING THE SENSES

To be successful a deceiver needs imagination and a sense of theatre. Deception during the American Civil War was by no means confined to the Confederacy. During Major-General U.S. Grant's Vicksburg campaign Major-General William T. Sherman was sent to make a demonstration up the Yazoo River, where his men were exhorted by their red-haired commander that every man was to 'look as numerous as possible'.[4] Even more theatrical was the trick attributed by Frontinus to the Athenian Pericles. Having notice a grove dedicated to Pluto and visible to both armies,

> he took a man of enormous stature, made imposing by high buckskins, purple robes and flowing hair, and placed him in the grove, mounted high on a chariot drawn by gleaming white horses. This man was instructed to drive forth, when the signal for battle should be given, to call Pericles' name, and to encourage him by declaring that the gods were lending their aid to the Athenians.[5]

The enemy promptly fled the field.

One memo sent to an American unit engaged in deception during the Second World War criticized the men's attitude for being 'too much MILITARY and not enough SHOWMANSHIP'; the writer went on to inform the soldiers that they must consider themselves 'a travelling show'. Presentations required 'the greatest accuracy and attention to detail. They will include the proper scenery, props, costumes, principals, extras, dialogue and sound effects. We must remember that we are playing to a very critical and attentive Radio, Ground and Aerial audience. They must all be convinced.' The author also admonished a colonel of *camoufleurs* who had instructed his men that all they need do was inflate their dummy tanks, after which they could go to sleep. 'This is very bad "theater". The Colonel forgot we were in show business and thought we were dealing with real tanks and real tankers . . . They must repair "Tanks", hang out washing, go looking for cider, and generally mill about in GI style.'[6]

The American Civil War provides another example of how corroboration via other senses such as sound and smell can be used to add credibility to a dummy position. In September 1863 Union general William S. Rosecrans wished to cross the Tennessee River without bloodshed and capture Chattanooga. A Southern officer later told a Northern correspondent that 'when your Dutch general Rosencranz [*sic*] commenced his forward movement for the capture of Chattanooga, we laughed him to scorn. We believed that the black brow of Lookout Mountain would frown him out of existence, and he would dash himself to pieces against the many and vast natural barriers that rise all around Chattanooga.' Instead, keeping his main body well back from the Tennessee River, Rosecrans demonstrated upstream; he ordered three brigades to light bonfires every night close to all possible crossings, and special details were instructed to chop and throw wood scraps into the tributaries, while others bashed away on empty barrels to imitate the sound of boat-building. On 21 August he added artillery attacks on the town itself and Rosecrans' opponent, Braxton Bragg, withdrew a brigade that had been guarding the area of Bridgeport some fifty miles downstream. A Union crossing was then immediately effected at three sites downstream including Bridgeport, where a pontoon bridge was built to replace the destroyed railway bridge. Rosecrans' entire force was across by 4 September, and on 8 September Chattanooga fell without a shot being fired.[7]

Smells, especially of cooking, are easily created, and if nowadays it is unlikely that realistic sounds can be reproduced by so simple an expedient as bashing barrels, nevertheless modern recording and amplification equipment present enormous possibilities. The sounds of battle – tank movement, bridge-building and so on – can be very effectively imitated, especially at night or in smoke-screens or where the enemy's aerial recce is weak. One place where sonic deception was developed by the British during the Second World War was Laggan House. There it was considered vital to be able to reproduce the sounds of specific types of tank, such as Shermans or Churchills, since experienced soldiers could be expected to spot the difference.[8] But sonic deception could also have a great psychological effect, particularly at night, as Geoffrey Barkas noted:

I see it as a purely emotional attack on the nerves. A sense of tension or fear is often built up in the minds of a cinema audience by arbitrary and illogical use of sound accompanying a picture ... Sounds by themselves are very frightening at night if they are associated with ideas that have caused the listener acute apprehension or suspense. In the circumstances the average listener does not stop to work out whether the sounds are strictly logical or accurate. His hair just stands naturally on end. I know mine does within the limited scope remaining to me ... It would be a most unusual enemy sentry or local commander who listened carefully and then said, 'All right boys, go back to bed. That noise is a General Grant [tank] and I know for sure that there are no General Grants within fifty miles.'[9]

DOUBLE AGENTS

Spies of one sort or another have for centuries proved a major source of information. Their efficiency and reliability depend on a great many variables. For a long time Prince Eugène of Savoy-Carignan had the postmaster at Versailles in his pay. The postmaster opened the letters and orders that the French court dispatched to its generals and sent copies to Eugène, who usually received them sooner than the French commanders. Double agents have long been used to convey false information to the enemy, as Frederick the Great wrote:

[François, Duc de] Luxembourg won over a secretary of the English king who informed him of everything that was going on. The king discovered him and turned the delicate affair to his advantage by forcing the traitor to write [to] Luxembourg and let him know that the allied army would make a large forage the following day. The French were nearly caught by surprise at Steinkirke [1692] and would have been entirely defeated if they had not fought with extraordinary valour.[10]

It is in the twentieth century, however, that double agents have made the biggest impact. John Masterman's *The Double Cross System* gives the impression that British deception of the enemy in the Second World War rested solely on the control of their agents and the feeding of false information through that channel. Masterman, however, was not the mastermind he seems to have thought himself to be. ('I ... had a more than average share of that moral and intellectual superiority which is ... the curse of the British liberals.'[11]) On the contrary, organized deception started in the Middle East in late 1940 as a normal military operation, in the course of which it was found convenient to make use of 'turned' agents deliberately to mislead the Italians, Germans and, later, the Japanese. The technique for using these agents in support of deception plans (as distinct from the more usual tasks of penetration and counter-espionage) was developed there by 'A' Force, as were all the other deception devices and arrangements later used so successfully worldwide by the British and the Americans.

While physical display may be ambiguous (it cannot, for example, disclose the commanders' names), electronic deception also has limitations. Apart from needing considerable resources and skill to disclose false information, there is no guarantee that the enemy will hear or is even listening. Double agents, however, combine the precision, certainty and speed necessary for deception at long range and over an extended period. But physical and electronic deception should be regarded as necessary security measures in such instances, since if the enemy does break the security ring, he will find nothing to contradict and, if possible, something to confirm the story.[12]

The use of double agents in the Second World War came about simultaneously in Britain and Egypt but was dealt with entirely separately in the two countries: in London by section B1a of MI5 under Lieutenant-Colonel T. A. Robertson, and in Cairo by Security and Intelligence Middle East (SIME), which was set up in April 1940 and under whose umbrella all matters pertaining to security and intelligence matters were combined. There Captain (later Colonel) W. J. Kenyon-Jones was chosen not for his academic qualifications but his business and athletic ones: he was managing director of Ronson's and a Welsh rugby international. He would in due course become deputy head of SIME. The head was directly answerable to Wavell as the commander-in-chief and when 'A' Force was created, although an operational unit, it utilized mainly intelligence channels. SIME was in fact a remarkable amalgamation of all security and intelligence affecting the Middle East, in which officers from MI5 and MI6 worked closely

CHEESE network including STEPHAN, an Austrian Jew who, as ULTRA decrypts later demonstrated, was the most highly regarded as far as the Germans were concerned. SIGINT also showed the British that there was no genuine Axis espionage network in the Middle East, although the Germans desperately wanted one, and as their ability to conduct aerial reconnaissance was reduced, so this desperation increased. In the mood of overconfidence inspired by the success of 1940 the Germans failed to build up a network in Syria or Iraq before the expulsion of Vichy forces or the Raschid Ali rebellion in Iraq in 1941. By the time of the Battle of El Alamein in October 1942 they were satisfied they had established in Egypt, through CHEESE and others, an espionage organization of supreme quality exactly as they had envisaged. They therefore acted with complete confidence and when the information supplied by this network was confirmed by wireless intercepts in the field, they found themselves quite, quite wrong, although any blame was attributed to camouflage by the British. This was the essence of 'A' Force operations.[14]

From the beginning of 1942, with deception becoming increasingly important, Dudley Clarke insisted that certain agents be allocated solely to deception purposes and the CHEESE network became that of 'A' Force. In the Middle East private armies were brought firmly under the General Staff and operated under the control of the commander-in-chief, unlike the situation in Britain, where the chain of command and means of control were not clearly delineated.[15] Clarke laid down certain principles for the employment of double agents: the contacts of an agent used for passing deception should be entirely notional, as should be his own espionage activities; a deception agent must not be allowed access to the outside world, irrespective of his own allegiances; and no deception link should ever be used for intelligence purposes other than deception. Should this occur for any reason, then the link should cease to be used for deception.[16] These principles were not followed in London and the results very nearly proved disastrous.

ELECTRONIC DECEPTION

Electronic deception (ED) can take two forms: physical (for example, deflectors and chaff to interfere with radar) or electromagnetic (such as false radio nets). The latter in turn can take three forms: imitation, manipulation or simulation. Imitation is by far the most difficult to achieve and the easiest to reveal, which makes it a dangerous proposition for the imitator. By appearing on the enemy's radio nets, issuing false orders for example, it is potentially very damaging. But if the radio procedures of the target are of a high standard and if the language skills of the imitator are not, imitation may prove counter-productive since, once discovered, the enemy is likely to take steps to improve his procedures, and this might drastically reduce the intelligence value of intercepts.

Manipulation and simulation are far more common. The former involves altering one's own electronic order of battle (the normal 'signature' of one's radio nets and procedures). By false traffic levels and controlled breaches of security the enemy may be denied a true picture of one's intentions and genuine order of battle. Thus manipulation contributes to security. Simulation, on the other hand, can be used to paint a completely false picture, to create electronically a false order of battle or inaccurate locations of a genuine order of battle. It will almost certainly be part of a larger scheme including the display of dummy equipment and perhaps 'special means' (double agents) used to corroborate the deception. Two important policy aspects are worth noting. The enemy cannot be given the designation of the formation by this means, nor can intentions be revealed, except insofar as these may be deduced from the type and grouping of formations and the character of training. In Britain during the Second World War dummy traffic was forbidden on the ground that the German cryptographers might deduce its nature from its pattern. The opposite view was held in the Mediterranean theatre, where it was believed that any increase of 'live' or genuine traffic added 'depth' to the cipher material and made it easier for the enemy to break it. Either way, live traffic certainly involves more and very highly skilled work. It also risks leakage in the traffic from a slip on the part of an operator, which cannot arise with dummy messages.[17]

PSYCHOLOGICAL OPERATIONS

Consideration must also be given to psychological operations (Psyops). There have been instances where Psyops have been enlisted to assist deception schemes and much Psyops hardware may also play a deceptive role:

loudspeakers, for example, may serve both functions, but there is a fundamental difference between the two. The underlying principle of good Psyops (and of propaganda in general) is that to be most effective, they must deal with the truth and nothing but the truth – although not necessarily the whole truth. Deception, of course, deals with lies, and its aims are therefore fundamentally at odds with those of Psyops. The latter can assist deception by helping to provide it with the 'bodyguard of truth' necessary to protect the lie, but those who practise either must be extremely careful that they do not compromise the other's position. For example, it must be remembered when designing Psyops materials, that no commander would ever allow operational details to be broadcast before execution and any intelligent enemy would know that.[18] Psyops will on occasion permit a lie, either in the form of 'black' propaganda purporting to come from a false (usually enemy) source, or if the lie is very credible and serves a specific aim. Credibility is one principle shared by deceivers and Psyops, since if a Psyops message contains an absolute truth that is not perceived as credible by the target audience, then it will probably fail.

Nevertheless, Psyops and the pattern of their use may act as intelligence indicators to an enemy. If leaflets are dropped in a particular place, for example, this may suggest that operations are being planned in that direction. It is obviously necessary therefore to co-ordinate Psyops and deception from the highest level. During preparations for the invasion of Sicily in 1943 (Operation HUSKY) deceivers looked to the Political Warfare Executive (as the

Psyops branch was then known) to assist them by controlling the numbers and locations of leaflets dropped that called upon the Germans to surrender. One area the deceivers wished to leaflet was the south of France in order to suggest forthcoming operations there, but aircraft were not available in the Mediterranean to do this and aircraft based in Britain could not reach that far. Therefore the threat had to be developed through radio propaganda.[19] This created severe problems since the Political Warfare Executive (PWE) had to be very careful not to raise false hopes among resisters; the plan also risked eroding the credibility of not only their own radio stations but crucially of the BBC, both of which broadcast to the whole of occupied Europe and both of whose credibility was absolutely fundamental to the success of their own operations.[20] As a general rule, it is probably safest for Psyops to expound one's capabilities while making no reference whatever to one's intentions.

SUBSTITUTION: EL ALAMEIN, 1942

The principle of substitution is to make a show of what one wishes the enemy to see (whether real or false) and then replace the item with something of a quite different nature and significance, according to the particular aim. The period between the battles of Alam Halfa and El Alamein during the autumn of 1942 was deemed essential for the regeneration and reinforcement of the forces that would be required to launch the offensive operations of the latter. The Axis forces, amounting to some 50,000 Germans and 54,000 Italians, were behind extensive minefields that, owing to the sea to the north and the Qattara Depression to the south, could not be outflanked. Although the Eighth Army under its new commander, Lieutenant-General Bernard Law Montgomery (soon known to all ranks as Monty), expected to have material superiority, at least on paper, some other way was needed to secure a break-through. The 'other way' was a deception plan on a scale never previously attempted. As the Eighth Army's chief of staff, Brigadier Francis de Guingand, told Dudley Clarke:

> Well, there it is. You must conceal 150,000 men with a thousand guns and a thousand tanks on a plain as flat and hard as a billiard table, and the Germans must not know anything about it, although they will be watching every movement, listening for every noise, charting every track ... You can't do it of course, but you've bloody well got to![21]

The overall plan for the battle was given the code-name LIGHTFOOT while the deception plan, written by Clarke, was called BERTRAM; subsidiary plans were code-named DIAMOND, BRIAN, MUNASSIB, MARTELLO and MURRAYFIELD. The intention was to conceal the huge build-up of forces in the northern sector of the Allied line where the attack was intended to be launched while simulating a huge build-up in the south; it was also intended to suggest that the attack would not take place until the end of the first week in November, when in reality it would begin on 23 October.[22] The GHQ Camouflage Section was given the task of concealing the huge dumps of rations, ammunition, fuel and stores that would be accumulated in the northern sector and making it appear that preparations were proceeding in the southern sector. This was done by disguising the real build-up as something more innocuous than it was and then inviting the enemy to look at it.

The first problem was concealing the huge quantity of stores necessary for so large an operation. Fortunately, it was discovered that there were a hundred sections of slit trench near El Alamein station, beautifully lined with masonry. Into these were put 2,000 tons of petrol, which air observers were invited to locate. They tried and failed. Food was delivered at night on 10-ton lorries and immediately stacked in the shape of 3-ton lorries and suitably camouflaged under nets, with any overflow stacked beside it in the shape of a driver's 'bivvie'. Similar methods were used for ammunition, engineer and ordnance stores. BRIAN was a scheme to create large fake supply dumps in the south, counterparts to those in the north. DIAMOND was a scheme to build a 20-mile fake water pipeline, ostensibly to supply the large 'build-up' in the southern sector. The trench to carry it was dug conventionally and fake railway line was laid in the trench to simulate the pipe. Before each stretch of the trench was filled in, the dummy pipe was removed at night for use further along the trench. Three dummy pump-houses and fake reservoirs were also constructed and traffic diverted and driven alongside it.[23] Significantly, it was built at a rate that suggested the D-Day of any subsequent operation would not be until at least ten days after the real one.[24]

In the southern sector there were real troop concentrations, including 44th (Home Counties) Division and 7th Armoured Division as well as 4th Armoured Brigade, which were detailed to carry out a diversionary attack. During the retreat to El Alamein 4th Armoured Brigade had suffered such severe casualties that it ceased to exist as an effective fighting formation. It quickly 'reappeared', under command of Captain Victor Jones of 'A' Force (promoted temporary major and local brigadier) and equipped with dummies, with the task of menacing Rommel's southern flank. By the time it was revealed through ULTRA that it had been rumbled, it had reformed as 4th Light Armoured Brigade, with the armoured cars of 11th Hussars and some Stuart light tanks operated by 12th Royal Lancers and the combined 4th/8th Hussars, plus a brigade's worth of dummies operated by a squadron of 3rd County of London Yeomanry.[25] Thus, during the Battle of Alam Haifa, when the Germans' armoured recce rounded a British minefield, they were fired on with real shells by a formation that according to their maps was a dummy one.[26] At the same time, the diversionary attack allowed the double agents who were supporting the deception plan to avoid accusations of having deliberately sent false reports.

MARTELLO and MURRAYFIELD were complementary plans that formed the most important aspects of the overall scheme, enabling the stationing of the armour of X Corps in assembly areas close to its start lines. Thanks to

Martello, by 6 October about 4,000 genuine vehicles, 450 dummies and more than 700 'sunshields' were in place. These – originally an idea of Wavell's – were covers that fitted over tanks to make them appear from the air like lorries. Meanwhile a force equivalent to X Corps was concentrated in an assembly area near El Imayid, to accustom the enemy to their presence.

On 6 October there was a change to the genuine plan Lightfoot. Naturally this had a knock-on effect on Bertram. Originally, the Martello area was expected to be some fifteen to twenty miles south of a series of tracks that led both south and north. The new area could only portend an attack in the north, and an additional deception was required. Three staging areas, Murrayfield (North), Murrayfield (South) and Meltingpot, were established astride a series of tracks tending towards the south. The armour was moved into these staging areas quite openly between 19 and 21 October. Just before the opening of the battle they were moved into Martello under their 'sunshields'. This presented the Camouflage Section with a difficult and unforeseen problem, as there would now be gaps left by the tanks moving forward. But Captain John Baker from the Camouflage Development Centre at Helwan, an architect in civilian life, made 'tanks' from the plaited panels of split palm that the local farm workers used as beds. In a remarkable piece of improvisation thousands of these panels were made by local workmen and knocked together in rudimentary form by three pioneer companies (one East African, one Mauritian and one from the Seychelles) together with No. 1 Camouflage

Company, Royal Engineers. Obscured under camouflage nets, they were all that was necessary to create three large 'armoured formations', apparently camouflaged and awaiting movement orders, probably to the south.[27]

Similar conjuring tricks hid their supporting artillery. Plan MUNASSIB involved digging gunpits with dummy guns at the eastern end of the Munassib Depression (south of the sector where the main attack would come) to represent three and a half field regiments. They were left without any sign of normal movement around them in order to convince the Germans that they were dummies. Shortly before the attack was due to commence, real guns moved in and joined the assault, a ploy that was also intended to reinforce the German belief that the main assault was still due to come in the south. Much of the supporting artillery was hidden in this fashion. A 25-pounder and its limber would be hidden under covers called 'cannibals' (also designed to resemble lorries), as were the distinctive Quad tractors. No fewer than 360 guns were concealed in this way and ready to launch the attack on 23 October.[28] To bring them forward, the Royal Engineers had to bulldoze tracks from MARTELLO through cannibals 1 and 2 to the frontline. This had to be organized so that they followed the least conspicuous course, and began from different places so as not to appear as a coherent track scheme until the last minute.[29]

To conceal that the guns and tanks had moved from the rear was the purpose of Plan MURRAYFIELD. First Armoured Division and 74th Armoured Brigade (Dummy Tanks) moved forward from Wadi Natrum to El

Imayid in two stages. The first was carried out openly as a training exercise to a staging area south of Burg-el-Arab, before moving to the forward positions at night with immediate replacement by dummies. This involved 1,500 vehicles from 2nd New Zealand Division, 1,370 dummy trucks, 64 dummy guns and 30 dummy tanks occupying the space vacated, although to aerial reconnaissance no move had apparently occurred.[30] Plan MELTINGPOT saw 10th Armoured Division move from Wadi Natrum by day to a staging area far to the south, then return at night to the main assembly area in the north, having left behind a mixture of dummy and real equipment. Another kind of display was made for the first time at El Alamein by deploying rafts out to sea between El Data and Sidi Abd el Rahman to simulate an amphibious assault, using a combination of noise and smell. Behind a smokescreen the reek of cordite and diesel was combined with confused shouting and the firing of flares. It was not especially successful, but was useful practice for future operations.

When General der Panzertruppen Wilhelm Ritter von Thoma, commanding the Deutches Afrikakorps (now just one component of Generalfeldmarschall Erwin Rommel's Panzerarmee Afrika), was captured, he said that German reconnaissance had failed to locate any increase in forces in the north, only in the south. His statement was supported by other prisoners and by captured documents, including a map that showed the supposed position of three British armoured divisions as being where the fake concentrations of transport had been. Von Thoma confessed that he had been certain that the assault would come from the

Munassib region, to the extent that two Axis armoured divisions had been retained in that sector until four days after the launch of the real attack. Furthermore, Eighth Army had been able to deploy a complete armoured division entirely unknown to the Germans.[31]

The False Routine: Crossing the Canal, 1973

Repeated often enough, any process becomes a routine of little interest. Many escapes from prisoner-of-war camps during the Second World War took advantage of routine events, the entry and exit of service vehicles, visits of Red Cross officials and the departure from work of German officers. Sentries are only alerted by the unusual.[32] It was after lulling the Israelis into such a false sense of security that the Egyptians launched the Yom Kippur War in October 1973.

The Egyptian planners wanted to slow down the Israeli response and especially to prevent a pre-emptive Israeli strike before completion of their own build-up. The Arab countries had carefully studied the lessons of the Six-Day War. In particular, they paid close attention to intelligence, communications and deception, including those measures taken by the Israelis in 1967 and by the Allies before the invasion of Normandy in 1944. Some among the staff had served with Montgomery at El Alamein and drew upon that experience. The resulting deception plan was therefore a blend of Israeli and Western techniques. Most importantly, it cleverly capitalized on Israeli and Western perceptions of the Arabs themselves, including a perceived inability to keep secrets, military inefficiency and inability to plan and conduct any sort of co-ordinated action.

The Israeli concept for defence of the Suez Canal assumed a 48-hour warning period would be sufficient, since the Egyptians would not be able to cross the canal in strength and could be quickly and easily counter-attacked. For this purpose, the Egyptians carried out at least six studies of Israeli doctrine and perceptions, enabling them to mesh these together skilfully in their own deception plan.[33] The aim of the plan was to provide plausible – and incorrect – alternative interpretations for the massive build-up along the canal and the Golan Heights. Put simply, the strategy involved increasing the 'noise' that the Israelis had to contend with by a series of false alerts. Over the previous two years there had been a series of continual escalations and backings down, beginning in December 1971, when a limited attack across the Suez Canal was averted only because President Anwar Sadat believed the Indo-Pakistan War would draw world attention away from the Middle East. A year later a smaller operation was cancelled when the general in command insisted that his troops were not ready. The result was to establish an apparent pattern of unexplained escalation and rapid release of tension.

By Egyptian accounts there were three such alerts in 1973: in May, August and late September. Each of these was accompanied by bellicose rhetoric. The first developed out of the situation in Lebanon following an Israeli raid on Palestinian headquarters in April. Fighting broke out between the Palestinians, who blamed the Lebanese army for lax security, and this led to Syrian units standing to maximum alert. Egyptian newspapers were instructed to publish civil defence notices and other items to increase the temperature. The crisis resulted in a split within the Israeli hierarchy. The Minister of Defence and Chief of Staff both considered it serious enough to warrant a counter-mobilization while the Director of Military Intelligence did not. The Ministry of Defence ordered mobilization anyway and – perhaps as a result – the situation developed no further. This mobilization cost Israel $10 million and military intelligence felt its assessment had been vindicated.[34] The second alert added to this sense of justification, and by the time of the third the whole thing was passed off by the Israelis as no more than agitation designed for local consumption.[35]

While the constant raising of alerts appeared to be forming a pattern, the Egyptians developed further ploys to explain this behaviour. The theme was that the Arabs, and particularly the Egyptians, were incapable of fighting a war and preferred to work towards some sort of diplomatic solution. This suggested that the alerts were mere sabre-rattling to create pressure and to placate hawkish opinion at home. This in turn played on Western and Israeli belief in Arab military incompetence, which the Arabs further enhanced by putting out the story that, following the expulsion of Soviet advisors in July 1972, Soviet-supplied equipment was rapidly deteriorating and the Soviets had expressed dissatisfaction with the level of Egyptian training (which was true). Following the

Israeli-Syrian air battle of September 1973 Syrian dissatisfaction with the performance of Soviet equipment
was made plain and the freedom of movement of Soviet
advisors was severely curtailed (possibly as a security
precaution to protect the invasion plans) amid talk of an
Egyptian-style expulsion. At the same time measures were
taken to suggest that Sadat intended to take Egypt's case
to the United Nations: he told a European foreign minister in the strictest confidence that he would be there in
October, knowing this would be passed on to the Israelis.
These measures culminated in the Fourth Non-Aligned
Conference in Algiers during the first week of September,
which passed a resolution calling for Israeli withdrawal
from Sinai. A visit by the Romanian defence minister to
Cairo was announced and an important speech by Sadat
was scheduled for 18 October, all of which was designed to
suggest growing impetus towards a diplomatic solution.[36]

Thus the stage was set for the final phase, the critical
period immediately before the attack, when Israeli action
could have proved catastrophic for the Egyptian cause.
The biggest problem was 'explaining' the unmistakable
and unconcealable movement of troops and equipment
towards the borders. Following the Six- Day War of 1967
the Egyptians had begun to fortify their side of the Suez
Canal, a process that, although speeded up from 1972,
was still slow enough to conceal its true potential from the
Israelis. In late September troops and heavy equipment
began to arrive under the cover of the annual autumn
exercises, a common phenomenon. Confusion remained
the hallmark, however. Ammunition was conspicuous by

its apparent absence, but had actually been sent previously by rail at the time of the May crisis and concealed in underground storage facilities. Troops were moved forward by day and appeared to return to their barracks at night, but only half the units would in fact return, thus concealing the build-up. Taking a leaf out of the Israelis' book, special bridging equipment was brought forward in crates to hide its significance. Specialist equipment such as water cannon (with which to blast Israeli sand ramparts – a technique developed during the construction of the Aswan Dam) was brought forward only at the last minute.

The 'exercises' were conspicuously highlighted in the papers, another fact suggesting they were designed for domestic political consumption. At the same time radio traffic was blatant during the day, and then when the 'exercises' ended, communications were carried on by a secure landline system previously installed. Reservists were called up, but in another imitation of an Israeli ploy they were also issued demobilization orders. A further set of orders was issued allowing troops to go on a minor pilgrimage to mark Ramadan. Special 'lazy squads' sat on the canal bank fishing, dangling their feet in the water and eating oranges, adding to the general air of unconcern (and military inefficiency) that they wished to convey to the Israelis. Security was tight, with operational orders not being issued to the lower formations and individual units until forty-eight hours before the launch. The Israelis later discovered that more than eighty- five per cent of Egyptians captured had no prior knowledge of the plan. One soldier said to his platoon commander as they paddled across the canal: 'So, 'construction training for special purposes', was the basis for the famous Brandenburgers, who were later expanded to divisional size, and destroyed after being reduced to fighting partisans in Russia. But not before it had been through some famous exploits.[39]

After operations in Scandinavia in April 1940 the Brandenburgers were faced with critical missions in the plan for the invasion of the Low Countries the following month. Over the previous winter and spring the Germans had spread a mood of insecurity and confusion throughout Europe through rumours of a large and highly organized 'fifth column'* of clandestine operatives deployed all across the countries they proposed to conquer. Although these were untrue, the belief helped stimulate panic in the rear areas during the campaign of May and June 1940, causing enormous problems for the Allies.[40] At a press conference on 21 May 1940 the Dutch Foreign Minister claimed that the Germans had dropped parachutists on Rotterdam and The Hague dressed as nuns, monks, nurses and tramcar conductors. It did not occur to anyone that these were patently silly disguises for troops engaged in an open invasion.[41] In fact, the German forces assigned to conquering the Netherlands were relatively weak and based on airborne forces landing in and around Rotterdam and The Hague. Two troop trains were to be rushed across to reinforce them and would be followed by the sole pan-

---

* The term originated during the Spanish Civil War, when the Nationalist general Emilio Mola made a radio broadcast saying that the Nationalists were advancing on Madrid with four columns and claiming to have a 'fifth column' in the city itself.

zer division assigned to this part of the operation. This made the capture of the railway bridge at Gennep, where it crosses the River Maas from Goch heading towards the western Netherlands, of particular importance.

Half an hour before midnight on 9 May, a group of Brandenburgers slipped across the frontier disguised as Dutch military policemen. They were led by a Dutch-speaking corporal to the road linking the villages of Heien and Gennep and into the marshes beyond. At dawn the two troops approached the bridge, and as a genuine Dutch military policemen tried to telephone a message that Gennep railway station was under attack, a group of six men approached the eastern end of the bridge, two dressed in Dutch uniform, the others in raincoats. These men – all Brandenburgers – overpowered the Dutch guards. It was now time to seize the western end of the bridge. A telephone call was made to the western guard-room with the message that two military policemen and four prisoners were to be brought over. The four 'prisoners' were handed over in the middle of the 400-metre bridge. When news of the approach of the German troop trains eventually came through, some Dutch troops on the west bank tried to set up a defence with a solitary gun, but it jammed after the first round and the 'prisoners' now overpowered their guards with concealed weapons, not found in their perfunctory search when handed over. Caught between the two fires, the remaining bridge guard surrendered, and soon the men of the German 481st Regiment were speeding their way through the Peel defence line to relieve the hard-pressed parachutists in Rotterdam.[42]

A particularly well-known deception involving disguise was wrought by the Germans during their Ardennes
offensive in December 1944, Operation WATCH ON THE
RHINE. Special teams of soldiers from 150th Panzer
Brigade, commanded by the notorious Otto Skorzeny,
would be sent ahead of the main columns, dressed in
American uniforms and mounted on captured American
vehicles to seize bridges and sow confusion among the
Americans in Operation GRIFFIN. They would plant
rumours and issue false orders, change signs and mark-
ers and remove minefield signs and use them to mark
false fields.[43] Volunteers were called for who could speak
English, with emphasis on colloquial American. However,
few who turned up at Grafenwöhr for training had anything more than the most basic knowledge of the language. Similarly, Skorzeny's request for uniforms and
equipment yielded only a fraction of his requirements.
The volunteers were taught how to lounge around with
their hands in their pockets, how to chew gum, how to
light their cigarettes like GIs, and to reply to a challenge
with: 'Go lay a fucking egg!' but ideas of any more practical training were soon written off. Nevertheless, organized
mostly in four-man teams known as *Stielau* (named after
their commander) and operating in Jeeps, they followed
up the offensive that opened on 16 December 1944 and
swiftly penetrated the crumbling American front.[44]

In material terms they achieved little: of nine *Stielau*
teams, two were intercepted quickly and one actually
reached the River Meuse; they turned a few road signs
around, bluffed one unit into leaving a village it held, cut

previous attempts to take Gaza, in March and April, had failed bloodily. Allenby therefore decided to adopt with minor modifications a plan developed by General Sir Philip Chetwode to change the point of attack from Gaza, in the north, to Beersheba, in the east. The difficulties, particularly the need for thorough logistic preparation, could not hope to be concealed, and hopes for success rested on misleading the Turks as to the timing, scale and direction of the attack.[47] The Turkish and German commanders already believed Gaza the most logical point of attack, since it followed the main line of communication, minimized water supply problems and offered the prospect of naval co-operation. Allenby's Intelligence Branch therefore submitted a plan that aimed to persuade the Turks that a third attack would be launched at Gaza, supported by a feint to the east at Beersheba – the opposite of the real plan. Preparations for this operation began two months before the intended date of attack, and Major Richard Meinertzhagen also began by preparing the 'ground' on which he would sow the seeds of deceit.

Meinertzhagen sent a letter and money to the Turkish spymaster, thanking him for his assistance, and ensured that it went to the Turkish intelligence staff. The spymaster was duly shot as a traitor without even being interrogated. Having thus prepared the Turks for the story he wished to tell, and after a number of unsuccessful attempts, Meinertzhagen rode out into the desert supposedly on a personal reconnaissance near Beersheba until he encountered a Turkish mounted patrol, which chased and fired at him. Pretending to be wounded, he dropped a number of articles in the path of the Turks, including a pair of binoculars, a water bottle, and a haversack previously smeared with blood. This was the real core of the deception, since within the haversack were staff papers that implied an attack against Gaza with Beersheba as a feint, a notebook detailing transport, water and supply difficulties regarding the Beersheba area, and personal letters from officers stationed around Beersheba suggesting that it was a mistake to choose to attack at Gaza.

Meinertzhagen was fairly sure that information planted in this way would be rejected by the Germans and Turks and so devised various forms of corroboration. Included in the haversack was a cipher book to enable the Turks to read British wireless traffic from Egypt. Urgent instructions relating to the lost haversack were sent, including instructions for Meinertzhagen to report to GHQ for a court of inquiry, and warning him to return in time for the attack on 19 November. An Army news sheet carried on unobtrusive 'lost' notice and routine unit orders alerting the rank and file to be on the lookout for the haversack, and a copy of the news sheet was 'mislaid' into Turkish hands. Searches were made in the vicinity, which led to unplanned corroboration when two British soldiers were captured by the Turks. Unaware that it was a deception, they honestly (and therefore credibly) informed their captors that the British headquarters regarded it as a disaster.[48] Additional measures were also put in place. Approximately once a fortnight throughout the summer a cavalry reconnaissance would be made towards the Turkish positions at Beersheba, partly aimed at lulling them into expecting only demonstrations in that area. The logistic preparations had to be conducted in great secrecy at night and carefully camouflaged. These elements were then given over to visible naval preparations and the registering of artillery targets in the Gaza area on 27 October. Camps near Gaza were left standing and lit at night while the troops moving eastward were carefully hidden.[49]

Although captured Turkish documents show that British claims for the deception were somewhat exaggerated (the Turks correctly identified many of the Allied changes and the general move eastwards, and discounted the naval threat behind Gaza), the British nevertheless scored a notable success. Immediately before the attack they saw significantly increased defence work in front of Gaza and reduced work in front of Beersheba. Even after the attack commenced on 31 October the German commander, Kress von Kressenstein, failed to believe the reports of its weight and refused to send reinforcements.[50] The main assault was entirely successful, capturing important wells and outflanking the Gaza defences, which the Turks were forced to abandon in a hurry. Allenby entered Jerusalem on 9 December, thus fulfilling a promise made to the prime minister, David Lloyd George, of Jerusalem as a Christmas present for the British people. At the same time he fulfilled an ancient Arab prophecy that a great Deliverer, named the Prophet of God, would one day deliver them from the yoke of Turkish rule. After all, the British commander's name could be translated as *Allah-en-Nebi* ('Prophet of God'); certainly the Arabs liked to think so.

THE PIECE OF BAD LUCK:
'THE MAN WHO NEVER WAS'

The Piece of Bad Luck stratagem is a variation on the Unintentional Mistake insofar as it aims to suggest that a piece of information has been lost not through incompetence but by accident or mischance. One example of the strategem, Operation MINCEMEAT, is among the most famous of all deception operations, but it was in fact only part of a much larger plan code-named BARCLAY, which was designed to convince the Germans that the next Allied objective following the North African Campaign in 1943 would be the Balkans, rather than Sicily. The credit for the part played in BARCLAY by MINCEMEAT must be shared by Lieutenant-Commander the Hon. Ewen Montagu RNVR (who wrote the famous account of the story, *The Man Who Never Was*) and Flight-Lieutenant Charles Cholmondley RAF of Bla.[51] The idea arose from the crash on 29 September 1942 of a British aircraft off the coast of Spain. A body had been washed ashore carrying documents which the Spanish authorities had made available to the Abwehr. The documents were unimportant, but it occurred to Cholmondley that a deception might be wrought by similar means. Having secured the body of a young man in his thirties who had died of pneumonia, the distinguished London pathologist Sir Bernard Spilsbury assured Montagu that a Spanish *post mortem* would not detect that the man had not died following an aircraft accident at sea. When it was pointed out that the body was not and never had been very fit, Montagu assured a senior officer that 'he does not have to look like an officer – only a staff officer.'[52]

Operation MINCEMEAT was now set in motion. The Hydrographer of the Navy was consulted and Huelva chosen as the spot best suited for currents to deliver the body; the Germans were also known to have a highly competent vice-consul there. The body was given the notional identity of Major Martin, Royal Marines, 'serving' on the staff of the Chief of Combined Operations, Vice- Admiral Lord Louis Mountbatten. His identity was carefully crafted. He was provided with a briefcase containing three documents, including letters from Mountbatten justifying his making the journey in person and a Combined Operations special pass. He was also given personal items such as a letter from his 'father', which Montagu described as so redolent of Edwardian pomposity that nobody could have invented it.[53] But crucially, if the German High Command was to be persuaded to act, then it would need to be provided with a document at a sufficiently high-level passing between officers who would unquestionably know the Allies' true plans. This was to be a letter from the Vice-Chief of the Imperial General Staff, General Sir Archibald Nye, to the commander of 18th Army Group, General Sir Harold Alexander. It was carefully drafted in 'old boy' style, and finally approved on 13 April 1943. It indicated that the Allies were planning an attack on Greece under the code-name HUSKY (the code-name for the genuine operation planned against Sicily). Playing on Hitler's congenital obsession with the Balkans, it used the Dodecanese as cover, with General Sir Henry Maitland Wilson as the nominated commander. The Dodecanese had already been built up as cover for previous operations

THE LURE: 8TH TACTICAL FIGHTER WING, USAF, 1967

The lure is an ancient tactic particularly suited to creating ambushes. Although closely associated with irregular or guerrilla warfare, it is by no means restricted to it. As the United States Air Force (USAF) demonstrated, it can be applied to the higher reaches of technology. When the Vietnam War intensified in 1965, it became necessary for the USAF to develop electronic counter-measure (ECM) equipment and evasive flying tactics to counter the Soviet-supplied Fansong radars, which acquired targets for the SA-2 Guideline surface-to-air missile (SAM). The most significant development from that conflict proved to be specialized airborne ECM units known as 'Wild Weasels'.

The F-4 Phantom fighters of 8th Tactical Fighter Wing – the 'Wolfpack' – were deployed to Thailand in December 1965 as part of the USAF's ROLLING THUNDER bombing campaign, initially to provide escort to the bomb-laden and vulnerable Republic F-105 'Thuds'. By the following September they had enjoyed considerable success against North Vietnamese MiG-17s, but the North Vietnamese began to deploy MiG-21s equipped with Atoll air-to-air missiles, which posed a considerably greater threat. A rising toll was being extracted on the bombers (the 'Thud' was obsolescent) and Phantoms had to pull out of the programme to provide combat air patrols. The entire ROLLING THUNDER programme was very much a political one, instigated by President Lyndon B. Johnson, and despite repeated USAF requests to be allowed to attack the North Vietnamese airfields near Hanoi and Haiphong, these remained out of bounds to American bombing attacks. Therefore the only opportunity to meet them was in the air.

In December 1966 the commanding officer of 8th Tactical Fighter Wing, Colonel Robin Olds, was called to Headquarters Seventh Air Force to discuss the problem. It was decided to conduct an offensive fighter sweep to bring the North Vietnamese fighters to battle under conditions favourable to the Americans. This plan, named Operation BOLO, would be carried out by using a strong force of Phantoms to simulate a bombing mission of F-105 'Thuds' and lure the enemy into intercepting them. The Phantoms would then rely on their superior avionics and radar or infra-red guided missiles to engage the enemy, avoiding if at all possible the close-in dogfighting that was the MiGs' speciality. For the raid to appear genuine on enemy radars, the Phantoms had to adopt the same speeds, altitudes, approach routes, tanker rendezvous points and refuelling heights over Laos as a real bombing mission.[56] To enhance the effect, the bogus strike force, known as West Force, adopted the same call signs and communications procedures as the 'Thuds' and was fitted with ECM pods, which at the time only the F-105s carried. Once the MiGs had been drawn into a fight, their normal escape route into China would be blocked by another fighter formation, known as East Force, provided by 366th TFW. The 'Time on Target' was carefully planned so that the Phantoms would arrive in waves every five minutes for almost an hour over the enemy airfields, which was the MiGs' estimated endurance and would give them no respite. Further support came from tankers, specialist ECM aircraft and other fighters over Laos.[57]

The operation was complicated, but although the scheduled day – 2 January – provided only poor weather conditions that would hamper the Americans and give cover to the North Vietnamese over their airfields, it was decided to proceed. Olds led his flight over Phuc Yen at around 1500 hours but the North Vietnamese were initially slow to react. It took two major sweeps before a radar contact was confirmed and this aircraft refused to venture through the cloud cover. A third sweep precipitated the battle, however, by which time a second Wolfpack flight had arrived under Colonel Daniel 'Chappie' James, who called urgently: 'Olds, you have MiGs at your six o'clock!' Breaking to his left, Olds allowed his third and fourth flight crews to deal with this threat while he pursued a MiG that popped out of the cloud at his eleven o'clock position. It immediately retreated, followed by two air-to air missiles. Then another MiG appeared further to his left about a mile and a half away, which he caught and downed with a Sidewinder heat-seeking missile. Meanwhile, Olds' number two had destroyed a second MiG and his number three another, roughly simultaneously, while one of James' flight accounted for a fourth.

By now the sky was filled with wheeling jets as well as the additional danger of SA-2 missiles. Olds and James were now running short of fuel and turned to disengage, leaving the battle to Captain John B. Stone's flight. Stone arrived over Phuc Yen at approximately 1510 hours with three other Phantoms and these managed to shoot down another three MiGs before turning for home. In under a quarter of an hour seven MiG-21s (amounting to nearly half the total possessed by North Vietnamese Air Force at the time) had been destroyed for no loss, a feat made all the more remarkable because only three out of fourteen flights had been able to engage in the cloudy conditions.[58] Another small success was gained four days later by simulating weather reconnaissance aircraft, but thereafter the USAF had to rely on more standard escort and combat air patrol tactics to bring the North Vietnamese to battle.

THE DOUBLE BLUFF:
HANNIBAL AND THE BURNING BRANDS

Double bluff can be regarded as either the deceiver's salvation or his nightmare.[59] To pull off a double bluff the deceiver must be very confident, or perhaps desperate. There is no more difficult deception, and in order to succeed, it probably requires more accurate and detailed intelligence of the target's mindset than any other. But if successful, the deceiver can reap the rewards while the deceived is busy congratulating himself on his cleverness at spotting a deception.

Having suffered terrible defeat at the hands of Hannibal at the River Trebbia and Lake Trasimene, the Romans invested the powers of Dictator in Quintus Maximus Fabius, a shrewd and careful general who was nicknamed 'the Delayer'. Fabius found Hannibal in Apulia and camped six miles away, keeping the Carthaginian at arm's length. Fabius had the advantage of a larger army with ample provisions, while Hannibal's forces were dispersed in foraging parties. Fabius was criticized for not attacking, but part of his aim was to overawe Rome's allies, who were wavering in their commitment to her. The Apulians remained loyal as a consequence and Hannibal was forced to cross the Apennines towards the Bay of Naples, shadowed all the time by Fabius, who kept to the high ground and refused to be drawn into action, either by the exhortations of his subordinates or by Hannibal's efforts to provoke him.

Eventually, they reached the heights of Mount Massicus and, as Livy describes it, looked down upon 'the most delightful country in Italy ... being consumed by fire'.[60] Although his subordinates were horrified by his reluctance to attack Hannibal, Fabius remained unmoved. He had acted very cleverly since Hannibal could not hope to spend the winter in Campania with no towns under his control, and he would have to retreat across the mountains the way he had come, taking with him the vast quantities of plundered food, including thousands of cattle. Having detailed 4,000 men to hold the pass, Fabius encamped his main force on a nearby hill overlooking the approach road and within reach of the important Latin Way.

One evening Fabius was aroused from his slumber to see thousands of lights on the hillsides, apparently heading for an escape through the hills beside the pass. Minucius, the Master of Horse and second in command, demanded that the army be roused and moved to the foot of the pass to block Hannibal's escape. Fabius suspected that it was another Punic trick and refused to be drawn, saying that the force stationed to guard the pass would deal with this movement, as the lights moved up and over the crest. Fabius was right: it was a trick, but one that was intended to be seen as a trick and Fabius was its victim.[61] The troops guarding the pass, however, thought the Carthaginians had found some way to outflank them and rushed up the hill to find the 'escaping Carthaginians' were actually some 2,000 oxen selected from the vast herd of looted cattle, with burning brands of wood and dried grass tied to their horns, and which Hannibal's pioneers had driven up the slope. As the Romans milled around among the cattle, they were suddenly attacked by a small force of Carthaginian light pikemen who took advantage of the shadows to cut them down, now leaderless and unable to operate in their usual close formation. A surge of the cattle then separated them and they waited for the dawn, uncertain of what was happening. Down in the now unguarded pass, Hannibal and his army, loaded with the spoils of Campania, marched unmolested and camped on the far side. At first light Hannibal dispatched a contingent of Spanish mountain troops who fell upon the Romans, killing about 1,000 before extracting themselves and the pikemen.[62] Not only had this ruse drawn off the guard at the pass, but Hannibal knew his opponent: where his previous Roman adversaries had been rash, headstrong and impetuous, Fabius was cautious and canny. Hannibal was confident that Fabius would refuse to be drawn at the one time when it was necessary to act promptly.

CHAPTER FIVE

TACTICAL AND OPERATIONAL DECEPTION

MALCOLM: Let every soldier hew him down a
bough, And bear't before him; thereby shall we
shadow The numbers of our host and make discovery Err in report of us.

Macbeth, Act V, Sc. 4

TACTICAL DECEPTION IS very often the product of necessity and quick thinking. At the Battle of Brandywine
in 1777 the 15th Regiment of Foot earned the nickname 'the Snappers': they consigned their small supply of
ball ammunition to the best shots while the rest of the unit
ran from tree to tree, 'snapping' at the enemy with blank
charges of powder until reinforcements arrived.[1] On the
Tunisian Front in 1943 an uncoded German radio message was intercepted calling for air support on a British
position at a given time, and stating that the target would
be indicated with coloured smoke. British artillery was
registered on a nearby German strongpoint and when the
Stuka dive-bombers arrived on time, it was this that they
found ringed with smoke. The Stuka attack proved very
effective.[2]

At the very lowest level of war the soldier should be a good shot and a bad target. The prime example is the sniper, whose art came of age during the First World War. Quite apart from requiring superb skill at arms, the sniper is also trained to the highest degree in fieldcraft – the ability to move and operate without being seen. Individual camouflage is taken to its limit with such items as the 'ghillie suit' a camouflage suit that covers the whole body. Much of the sniper's work involves detailed observation and the collection of information. During the First World War the Lovat Scouts (Sharpshooters) were raised from among the Highland gamekeepers (ghillies) of Lord Lovat's Scottish estates. Originally intended as marksmen, their observational skills were such that they were employed to watch and report on German positions and troop movements deep behind the front lines.[3] However, snipers have always been regarded with ambivalence by other soldiers, who are sometimes happy to see snipers killed even when from their own side. Naturally, enemy snipers are prime targets and at this lowest and most personal level of warfare, deception plays a considerable part.

A common response to snipers is the creation of decoys to draw their fire. The Special Works Park made papier mâché heads for this task, but such refinements are not always available. Sergeant Quinn of 1st Airborne Division Reconnaissance Squadron, defending the Oosterbeek perimeter during the Battle of Arnhem in September 1944, constructed a decoy out of a pillow, a steel helmet and a broomstick, and exposed it at various places and at different times along his unit's section of the perimeter. It never failed to provoke a German sniper into giving away his position and its skilful employment accounted, it was claimed, for seventeen or eighteen of the enemy.[4] In 1917 the Special Works Park even produced a set of life-size dummy figures out of plywood, which could be raised on top of the trench parapet in twos and threes as required and dropped again, very much like targets on a range. Seen through the smoke and dust of battle they could create the illusion of an attack being launched. The figures were used at Messines Ridge in June 1917, by 46th (North Midland) Division, as distractions in so-called 'Chinese attacks', and again during the breakthrough at Cambrai the following November.[5] The same tactic was used by 9th Australian Division at El Alamein in 1942.

ATTACK: LE PLESSIS GRIMAULT

With modern weapons making camouflage and concealment essential, deception has come to permeate every phase of war at the tactical level. A deception may need to work for only a few minutes in order to give one side a crucial advantage, and the complicated intelligence process that involves signals intercept and aerial photography is unlikely to be of immediate concern to those personally engaged in fighting a battle.

Depending on the nature of the mission, a commander might choose to implement any number of measures to gain an advantage. In attack, for example, he might conduct covert reconnaissance of the real objective while making a more obvious recce of a diversionary one.

He might construct an obviously dummy bridge on the real approach route and conduct artillery target registration on the diversionary objective. Troops might move forward during the day and then be switched at night. Dust could be raised and a controlled breach of radio security might be permitted in a false concentration area, although at the tactical level things tend to be much simpler. The use of feints and demonstrations is very common since one is likely to be so close to the enemy position that these are likely to gain his attention. Smoke and artillery can be used to mask movement or indeed the lack of it, in order to draw the defenders' attention away from genuine preparations.

During the Normandy campaign in the Second World War, when American Third Army broke through at Avranches to pour first south and then east behind the German Seventh and Fifth Panzer Armies, Hitler's insistence there be no retreat helped to create the Falaise pocket, in which these German armies became trapped. The failure of the counter-attack at Mortain further drew the defenders into a sack and the two days beginning on 6 August sealed the fate of the Germans in Normandy. Driving south-east, American columns met with little resistance, but on the flanks of the Mortain salient a framework of hedgerows allowed the Germans to mount a skilful, coherent and stubborn defence. The commander of Fifth Panzer Army, General der Panzertruppen Hans Eberbach, finally abandoned to the British the ruined villages of Villers-Bocage, Aunay and Evrécy and the area north of Mont Pinçon, to form a new defensive line incorporating the mountain and still disputing every inch of ground.[6]

Mont Pinçon is the highest point in Normandy and formed a vital hinge in the German defences. Its capture was a brilliant feat of endurance by men of 43rd (Wessex) Division. At the foot of the hill to the south-west lies the village of La Varinière, which contains an important crossroads through which the assault troops had passed and which remained under heavy shellfire. On the southern side lies the village of Le Plessis Grimault, into which no fewer than seven roads feed including the main road from Aunay, along which 7th Armoured Division was pushing south towards Condé-sur-Noireau.[7] The 5th Bn, Duke of Cornwall's Light Infantry, was resting in reserve on the afternoon of 7 August 1944, four miles from Mont Pinçon with scouts and carriers forward attempting to locate the enemy. Late in the afternoon the commander of 214th Infantry Brigade, Brigadier Peter Essame, arrived to inform Lieutenant-Colonel George Taylor that he was to capture Le Plessis Grimault in order to open 7th Armoured Division's axis. They immediately set out to recce the objective before it became too dark and crawled to a position on top of Mont Pinçon to observe the village.

Taylor was determined not to repeat the frontal assault tactics that had recently cost his battalion twenty dead and sixty-nine wounded in a day's fighting at Jurques, when they had run into unexpected enemy armour. In his memoirs he recounts how detailed information about the enemy to his direct front was rarely available, and describes infantry tactics as being like those of 'a policeman tackling a desperate man in a dark room – one had to find him first with one hand before hitting him with a baton'.[8] Consequently, he devised a plan for a silent night attack by his A Company, closely supported by D Company and with C and B Companies held well back to provide a powerful reserve and maintain flexibility. Meanwhile, the attached B Squadron, 4th/7th Royal Dragoon Guards, would make a noisy demonstration from the direction of La Varinière. They were to be supported by a block barrage fired by three field and one medium regiment of artillery, together with mortars and medium machine-guns covering the right flank. The aim was to draw the enemy's fire, particularly that of his artillery and mortars, and when the tank squadron commander asked for a company of infantry in support, Taylor refused. Instead, he allotted a platoon since the armour would be relatively safe from indirect fire, but infantry in company strength would be very vulnerable to it.

Once the brigadier had accepted the plan and departed to coordinate the support, Taylor gave his orders to his company commanders' and as the light faded they moved through high bracken and birch trees over the crest of the hill. The artillery barrage commenced at 2130 hours and the tanks, with their supporting platoon from C Company well dispersed among them, advanced along the axis of the road from La Varinière and opened fire with their machine-guns. By now it was dark, with light signals being fired from the village and streams of tracer and heavy mortar fire being directed at the tanks, but an ammunition store then exploded. 'This magnificent

horrendously costly or even fail. In such circumstances deception becomes yet more important. The techniques are very similar to those of breaching a line (after Marlborough at the *non plus ultra),* where a demonstration or a feint may sufficiently weaken the defences at the chosen point to effect the crossing. Thus measures for obstacle crossing might include moving amphibious engineers up by day through a flanking formation and switching them laterally by night, and routing support and engineer traffic via deception routes by day, ostentatious reconnaissance of other crossing sites, simulation of crossings away from the actual site and use of a deception fireplan. In such circumstances use of airborne troops can be very effective in creating diversions. Indeed, judicious use of small forces in this way can be very effective in fixing the enemy and achieving a broader aim. The last exploit of 2nd Parachute Brigade during the Italian campaign took place on 1 June 1944, when three officers and fifty-seven men from 6th (Royal Welch) Bn, Parachute Regiment, were dropped an hour before last light with the task of preventing the Germans (who were in the process of withdrawing from the Pisa-Rimini Line) from carrying out large-scale demolitions. The drop was delivered perfectly on difficult ground and in daylight, and in order to disguise the small size of the force Brigadier C. H. V. Pritchard made extensive use of dummy parachutists. The Germans were tricked into believing a far greater force had actually been employed and as a result moved a regiment to deal with the perceived threat to their communications; another German division was forced to remain in its location rather than being brought forward to reinforce the troops in the frontline, which was then heavily engaged with Eighth Army.[10]

DEFENCE: THE EASTERN FRONT AND SOUTH-EAST ASIA DURING THE SECOND WORLD WAR

In defensive operations deception serves to blunt the enemy's attack by diffusing his energy onto false objectives and to create conditions for the defender to return to the offensive. The first aim is to divert enemy recce and electronic warfare effort towards unproductive areas through counter-surveillance measures, so that the enemy cannot identify key assets such as artillery and command and communications facilities. But deception can also protect key logistic assets and main supply routes, vital if one is later to return to the offensive. As far as defensive battle itself is concerned, deception aims to force the enemy onto ground of one's own choosing and to make the enemy deploy his forces, especially his artillery, prematurely. By concealing crucial moves – such as the direction and timing of counter-attacks – the defender can induce the enemy to deploy his reserves away from positions where they could influence the battle.

Passive measures to achieve surprise in defence are essential, if only to avoid the full effects of enemy artillery and air power. Avoiding obvious positions, particularly the front edges of villages and woods, and seeking the reverse slopes of hills and defilade positions (enabling the defender to bring fire to bear of an attacker's flank) should be combined with the careful camouflage and concealment of all vehicles, troop positions and stores (particularly, in modern warfare, against thermal imaging devices). Passive measures include the minimum use of radio and preferably the imposition of radio silence and other emission control measures: even today field telephones greatly enhance the security of communications. All movement, particularly of reconnaissance and work parties, must be planned with concealment in mind. A deception plan should always be considered to reinforce these passive measures, although it is likely to be limited by time restraints: before considering creating dummy positions a defensive position must be secured and, if possible, the troops dug in. To succeed, the plan must have a clearly established aim, which may include drawing enemy fire onto unoccupied areas or dummy positions and encouraging the enemy to waste time and ammunition (and to expose himself needlessly) by deploying to attack such positions. The simulation of a credible alternative to part or all of the defensive plan, for example by creating a false front, may encourage the enemy to react in this way.

Like all deception plans, those for a defensive position must be centrally co-ordinated. The over-riding principle should be to hide what is real and display what is false. Although major facilities and operational positions may need special camouflage stores and measures undertaken by engineers, it is otherwise surprising what a few men can achieve, even today. Trenches' need only be a foot deep provided the bottoms are lined with branches, and prove very effective if radar reflectors to recreate vehicles, heat and smell sources and a smattering of the usual impedimenta are left lying around. However, no dummy position is going to hold up an enemy forever unless he is very timid, like McClellan at Yorktown.

During the Second World War the Germans were never much impressed with the quality of the Soviet military system, but Soviet deception schemes went far beyond the concealment and deception inherent in the Germans' own camouflage techniques.[11] In the early stage of the war, when they were usually going forward, German troops operating on the Eastern Front often found it difficult to locate Soviet defensive positions. Patrols and vanguard units would be allowed to pass through apparently uninhabited regions only for defences of up to regimental strength to be revealed to the main body.[12] Friedrich von Mellenthin, a former German staff officer, wrote that:

> The Russian soldier is a past master of camouflage, of digging and shovelling, and of building earthworks. In an incredibly short time, he literally disappears into the ground, digging himself in and making instinctive use of the terrain to such an extent that his positions are almost impossible to locate ... Even after long and careful scanning, it is often impossible to detect his positions. One is advised to exercise extreme caution, even when the terrain is reputedly free of the enemy.[13]

Later in the war the Germans often found themselves responsible for defending vast frontages with very few

men. One difficulty was deciding whether to form a continuous line or concentrate in strongpoints. The strongpoints would afford closer control and greater strength
but leave dangerous gaps. A continuous line on the other
hand, would make Soviet infiltration more difficult and
reduce losses from concentrated artillery fire.[14]

Over the course of the war the Germans became
increasingly adept at defensive tactics. Partly because of
heavy manpower losses they retained a high proportion of
automatic and heavy weapons. It was therefore common
to hold an outpost line very thinly, to act as a 'tripwire' on
the attackers – making them deploy too soon and waste
artillery and other support on a thinly held crust, before
trying to throw the enemy back with counterattacks.
German machine-guns used to be continually switched
to alternative positions as soon as they had fired, partly
because this helped prevent their destruction by enemy
fire but also because it acted to deceive the enemy as to
the strength and true disposition of the defending force.[15]

Artillery batteries are vulnerable to counter-battery
fire but positioning in reverse slopes and among buildings
assists their ability to survive. The careful siting of radios
clear of the position (to hinder enemy direction-finding),
camouflage and concealment also help to prevent enemy
acquisition, as do false gun positions and ammunition
dumps. GHQ noted as long ago as 1918 that the real value
of dummy battery positions was not fully recognized: in
order to succeed, they must be planned very thoroughly
to present subtly all the indications of a genuine battery,
including blast marks and tracks. A dummy position

should be located as close to the real one as is consistent with security, and tracks may be used for both. If the dummy position is unsuccessful as such, it can still act as an alternative firing position.[16] 'Firing' can be simulated to deceive enemy sound locators by placing charges a metre off the ground. Flash simulators should also be used. Jasper Maskelyne learned how to make perfect flash simulators for 25- pounders in the Western Desert. Unfortunately, there were no scales to be found in that part of North Africa, so he and his men had to measure everything with spoons. Four teaspoons of blackpowder would create the smoke, six dessert spoons of aluminium powder the flash, and a teaspoon of iron filings the red flame. Differing quantities could be used for different-sized guns.[17] In order to retain credibility, dummies also need to move regularly and be mixed with some real guns so that at least a few rounds land at the other end.

In the jungles of Burma and on the coral atolls of the western Pacific, unlike the steppes of Ukraine or deserts of North Africa, visibility is very limited. During the Second World War an American combat patrol on Makin Atoll was pinned down by a Japanese sniper. An American scout worked his way forward to where he could see the enemy, crouching in a tree. Carefully, he took aim and fired. The sniper tumbled from the tree with limbs flailing. As the scout raised himself to signal his comrades, he was cut down by a burst of machine-gun fire. Later his friends found his target, made of clay and palm leaves, dressed in an old jacket and peaked cap and armed with a wooden pole.[18] The Japanese were so adept at such ruses that at first they completely baffled the Americans and British. Their ability to get behind blocking positions and create diversions of their own was especially effective in Malaya against the British, who were tied to the roads and many of whose troops were in fact poorly trained and unenthusiastic Indians. It was not long, however, before the strategic initiative passed to the Allies, first to the Americans in the Pacific following the Battle of Midway and later to the British in Burma, after two years' bitter fighting following an abortive attempt by the Japanese to invade India.

Many of the places where the Allies fought the Japanese consist of steep razor-backed ridges covered with dense forest. Here the opportunities for display were naturally very limited, but in defence Japanese infantry were expert at concealment. Their jackets often had a variety of loops and pockets sewn onto the reverse side to enabled grass or branches to be stuffed into them. They cunningly placed bunkers on the tops or reverse slopes of hills. These were immensely strong, constructed from logs and extremely well camouflaged with grass and moss, occasionally using nets, but seldom visible from more than fifty metres away. And they were equally adept at concealing important equipment such as aircraft, using a variety of decoys and other devices.[19] Similarly, dummy figures and weapons would be set up to draw fire and give false impressions of strength. On the other hand, the nature of the terrain provided good opportunities for sonic deception: the Japanese were very fond of shouting false orders in English or Urdu in order to persuade troops to surrender. Other ruses were brutally lethal. Dead and wounded troops would be used as bait to lure more men into the arcs of Japanese machine-guns, and extensive use was made of explosions on the rear and flanks of the enemy, using Chinese firecrackers if nothing else was available.[20] They were quite prepared to dig positions with the firing slits pointing backwards and to lie low while an attack moved past before taking it from the rear. Guile and deception were as much a part of everyday life in the Pacific and South-East Asian theatres as the mud, the heat and the insects.

RELIEF IN PLACE: THE HOOK, KOREA

If a unit or formation spends any length of time in a defensive position, it will need to be relieved at some point in order to avoid exhaustion, especially if it has suffered significant casualties. Relief in place is the measure by which all or part of a unit or formation occupying a piece of ground is replaced by another force who will adopt responsibility for it. However, the movements of two forces into and out of the same area is obviously a risky one, which the enemy might well seek to exploit and which provides him with a particularly good artillery target. Security is therefore essential, and deception measures can contribute greatly to it. Co-operation between the two defensive forces is paramount and command must be clearly delineated, especially the point at which the handover of command is actually made.

Deception may include the simulation of normal activity by continued radio traffic from the outgoing force, including the retention of operators from the outgoing force until the relief is complete. At the same time radio silence should be imposed on the incoming force. Normal administrative and patrol activity should be maintained, and patrols should also be conducted by the outgoing force until the relief is complete. The incoming force must restrict the size and movement of its advance party and only use the vehicles of the outgoing force; this is particularly important if the two are equipped with different vehicles. Artillery fire and deliberate vehicle noise might be used to cover any noise in the relief and measures need to be taken to conceal movement, including smoke and light to blind observers and surveillance devices. A demonstration on the flank using smoke, lights, artillery and vehicle movement might be made to indicate an imminent attack, reinforcement or withdrawal in an adjacent area but, since this is something that might in itself draw enemy attention, again co-operation is critical.

In the latter part of the Korean War some twenty divisions were fighting to hold the line of the 38th Parallel 'in defence of the principles of the Charter of the United Nations'. One of these was the Commonwealth Division, which was separated from the sea to the west only by a division of the United States Marine Corps. Two bastions of defence formed the cornerstones of the division's front: Point 355 and Hook Ridge. The five battles for possession of the Hook saw more British blood spilt and a greater concentration of enemy troops than any other point in the Korean peninsula. The second and costliest of these battles was fought by 1st Bn, Black Watch, in November

1952, and it was the Black Watch who once again held the feature on the prelude to the third battle. By this time, however, they were due for relief from the cumulative effects of casualties, patrolling and lack of sleep. The relief by 1st Bn, Duke of Wellington's Regiment, was scheduled for the night of 12/13 May 1953, and the handover was due for completion by 0700 hours on the 13th. Thorough planning, preparation, briefing and reconnaissance, combined with security and deception were vital to reduce the risks of this hazardous operation, which began as soon as Brigadier Joe Kendrew announced the relief on 9 May.

Fortunately, in the run-up to the relief, there was relatively little activity from enemy artillery and mortars while company and platoon commanders from the Dukes came forward to reconnoitre the positions they would be taking over. Understanding the complicated system of minefields to their front was another matter; in twenty months the positions had been occupied by as many units, each of which had added its own layer to the barrier. Little of the wire remained to mark the minefields and the Black Watch organized patrols to show the newcomers the gaps. (As the only access to no man's land, these were natural targets for ambushes and mortar fire.) For three days before the relief, the Dukes' signallers were on the position to learn the intricacies of the telephone system. Similarly, the Black Watch signallers remained on the position for three days after the move was completed to ensure that the voices on the radio were Scottish and not Yorkshire ones until the newcomers settled in. While conducting their recces, the Dukes were careful to ensure they wore the same badges

higher plan. Passive measures will include radio security, and no reference should be made to the withdrawal over any insecure radio net. Unless radio silence is already in force, every effort should be made to maintain the normal pattern until the withdrawal is complete.

Second, the maintenance of routine must suggest that nothing unusual is about to happen. Any established routine for harassing fire, artillery adjustment, vehicle and echelon movement, patrols and so forth, should be maintained. Active measures might include radio deception, but this will be conducted at formation level. Units might use artillery, mortar, tank fire and illumination to cover the noise of movement and to distract and blind the enemy, as well as improvising and simulating lights and noise to suggest the continued occupation of defences after the withdrawal is complete. Planning must ensure the careful timing of both preliminary demolitions and the destruction of unwanted stores and equipment.

The British expedition to the Dardanelles in 1915 has earned a reputation as one of the great disasters of the First World War. But despite the apparent incompetence of the High Command, the withdrawal was a brilliant feat; not a man was lost in an operation made infinitely more hazardous by being channelled into narrow beaches and waiting ships. When the expedition's land force commander, Sir Ian Hamilton, was relieved by Sir Charles Munro on 28 October 1915, Allied strength had already been reduced from a war establishment strength of 200,000 to 114,000. Hundreds of sick were being evacuated every week without replacement and the entire force was being steadily run down. Munro reported that the only realistic option left to the expedition was evacuation, a view that caused some consternation in London, especially when he added that he expected casualties in the region of thirty to forty per cent.[22] But with Lieutenant-General Sir William Birdswood appointed commander, the cabinet agreed on 7 December to evacuate Suvla Bay and ANZAC Cove, with a final decision on Cape Helles pending.

As with so many deceptions, it was not merely a fantastic piece of bluff but a supremely well-organized feat of logistical planning. The trick would be to persuade the Turks that nothing was happening by maintaining routine and manning the front-line trenches until the very last moment. The anchorages had to keep the pattern already set, so there could be no sudden arrival of transports in broad daylight. The gradual thinning out of artillery lines was covered by replacement with dummy guns and movement of the remaining real ones to alternative sites at night, something that required careful planning and practice. By the same token, there could be no apparent thinning out of stores dumps, medical units or daily administration. ANZAC presented the greatest danger because here the Turkish frontline was literally only a few yards away from that of the Australians and New Zealanders. At Russell's Top they needed only to move forward some 300 metres to dominate the cove completely with plunging fire.[23] In order to lull the Turks into accepting periods of total peace and quiet (so they would not notice the period when the Diggers were moving away), Major-General Sir Alexander Godley gave orders for the troops to observe periods of complete silence when not a single shot was to be fired. Should the Turks conclude that the Diggers had abandoned a position and moved into the open, they would be swiftly disabused of this notion by withering fire. The Turks soon decided that this was all just a bait and that the trenches opposite were as strongly held as ever. In reality, the ranks were being steadily thinned down to a skeleton force.

At Suvla the northern sector was joined to the bay by only a footbridge across a cut. Lieutenant-General Sir Julian Byng decided that this position would have to be evacuated separately and, in order to prevent Turkish interference from their nearby positions once this was discovered, intermediate lines of defence were first prepared halfway to the embarkation sites, although it was recognized that these and their defenders would have to be abandoned if the Turks pursued vigorously. The combined garrison of ANZAC-Suvla at this time was 85,000 men, 5,000 animals, 2,000 vehicles and some 200 guns together with vast stocks of ammunition, rations and other stores. By the middle of December a steady rearward flow was in progress, conducted at night according to a strict timetable under the supervision of the chief naval beachmaster. By the 18th the garrison had been halved but the normal artillery programme continued (with the remaining gunners firing twice as many rounds), snipers remained as busy as ever and barbed wire was continually renewed. Men were detailed to loiter around the rear areas and light fires in vacated medical tents while work parties fed the incinerators. Lighters continued to land boxes of 'stores' (filled with sand) and whatever rations that could not be removed or burnt were ruined and buried during darkness. Everything was done to preserve an air of routine. Even a small Turkish trench raid might have revealed the truth but the Turks had taken just as much of a beating in the preceding months of vicious fighting and were in no mood to take risks.

The soldiers entered into the spirit of deception fully by improvising a number of ingenious devices. The Aussies, for example, invented the delayed action rifle, by which fire could be sustained from a trench long after it had been vacated. A rifle was mounted on a parapet with a string tied to the trigger. Two kerosene tins were arranged above and below, which dripped water from top to bottom. When enough water had collected in the bottom tin, the string pulled the trigger. Others were rigged with fuses and candles, and the timing could be varied to produce a steady flow of 'sniper' fire for half an hour or more after the troops' departure.[24] Mining continued throughout December and January with notable successes; the last mines were scheduled to use up the final stocks of explosives and timed to detonate after the last troops had left the beaches.

This occurred on the night of 19/20 December 1915. By now only 5,000 men remained at ANZAC-Suvla to face 60,000 Turks. Suvla was evacuated undetected but with some anxiety. One of the piers was hit by enemy shelling during the final morning and had to be hurriedly repaired. But so carefully were the forward trenches vacated by 0130 hours that the Turks continued to fire at them all night. Mines were exploded and the night sky lit by the fires of burning stores dumps. None the less, the German advisors to the Turks reckoned enough stores were left to feed the Turkish troops for several weeks. At ANZAC there was intense competition among the Diggers to be part of the rear party. With the Germans sending heavy artillery to bolster the Turks, the evacuation was becoming increasingly urgent and enemy shelling was growing in intensity. On the final evening a large mine was prepared under the Turkish position of the Nek at ANZAC and the duckboards in the trench bottoms were pulled up to soften the footfalls of those departing. Barbed-wire barricades laced with booby traps were prepared, which the last man to leave, always an officer, would pull across to impede pursuit.

By 2200 hours only 1,500 picked soldiers remained. When volunteers had been called for, every man in a battalion had stepped forward: they were very reluctant to leave their dead mates behind. These last few began to withdraw at 0130 hours, taking the last machine-guns down to the beach, lighting the mine fuses and pulling the wire barricades across communication trenches. Lone Pine, the key point in the ANZAC defences, was the scene of a final minor triumph. After evacuation at 0240 hours a mine under the Turks in the Nek went up with a colossal roar at 0330 hours, as the last few men were being counted through the various checkpoints down to the beach. Only then were the final stores dumps fired. By 0400 hours there were no living defenders left at ANZAC. Deeply moved at the thought of leaving behind the graves they had so lovingly tended, one Aussie said to his officer: 'I hope *they* won't hear us marching off.'[25]

The success at ANZAC-Suvla greatly encouraged the planners at Cape Helles, although it was obvious that a repeat performance would be difficult to pull off. A debate continued on the merit of withdrawal; the Navy wanted to maintain the beachhead to assist its blockade of the Dardanelles. While this continued, an aggressive posture was maintained to persuade the Turks that the Army had every intention of staying put. Removal of unwanted stores, men and animals continued all the same. Reports from prisoners and deserters showed that the Turks were keeping a particularly close watch on the rear areas for signs of withdrawal and frontline units had been ordered to keep up the pressure. The bad weather was making embarkation difficult, and the sinking through collision of a transport earmarked for removing the large number of pack animals meant that mule transport units were faced with the sad task of destroying their charges. With the garrison down to 19,000 men and sixty-three guns, the Turks increased the pressure at the bidding of Liman von Sanders, their German commander, who was determined the British should neither stay nor leave. A heavy attack was launched on 7 January, accompanied by mines and especially heavy artillery support, but it was so thoroughly resisted that the Turks did not press it home. Seventh Bn, North Staffordshire Regiment, in particular, fought with such determination and tenacity that the Turks were convinced that evacuation was not being contemplated, yet it began that night.

By dawn the following morning 17,000 men remained ashore but the operation was continued as soon as it was dark following the same pattern as at ANZAC-Suvla. Feet were muffled by wrapping them in sandbags and shelling from the Asiatic shore continued as it had for months previously. A heavy swell at W Beach seemed to threaten the success of the operation, especially when a lighter was grounded at Gully Beach, compelling the men there to move along the coast road for W Beach with their commander, Major-General Maude, at their head. Suddenly, Maude realized he had left his luggage on the stranded lighter and insisted on returning for it. He made it back to W Beach triumphantly pushing a stretcher trolley with his bags, which were hurriedly bundled aboard a boat as the final fuses were lit. Once again a substantial force had made a withdrawal in the most difficult circumstances without suffering a single casualty. Once they realized they had been duped, the Turks sent up signal rockets and surged forward into the vacant trenches, only to be greeted by the hundreds of booby traps left for them.[26] According to von Sanders it was a master stroke and, given that the Turks held most of the high ground, it cannot be denied. Fortunately, somebody had also seen fit to blind those Turks on the vantage point of Gaba Tepe with a destroyer's searchlight, both on the night of the evacuation and for several nights previously.[27]

Gallipoli demonstrates how building up a mass of small tactical deceptions, each unit responsible for its own part of the plan, can enable a major operational-level deception to succeed. In the Second World War the British developed an organization in the Middle East that was at once unique and hugely innovative. Having begun with small displays of dummy equipment, they formed a unit to practise tactical deception, then learned how to add other instruments, providing the basis for successively greater deception schemes that in due course led to the ultimate deception covering the invasion of Normandy, Operation BODYGUARD.

DEVELOPING OPERATIONAL DECEPTION: 'A' FORCE

When Dudley Clarke died in 1974, his obituary recalled that Field Marshal Earl Alexander of Tunis had publicly stated that he had 'done as much to win the war as any other officer'. Such high praise from one so great – given that Clarke attained no higher rank than brigadier, was never knighted and remained unknown to all but a few of his contemporaries – prompted the obituary writer to conclude that Clarke was 'no ordinary man'.[28] If further proof of this were needed, Clarke's citation for the United States Legion of Merit originated in the White House and was personally signed by the president.[29] Dennis Wheatley described Clarke as 'a small man with fair hair and merry blue eyes, an excellent raconteur and great company in a party', but with 'an uncanny habit of suddenly appearing in a room without anyone having noticed him enter it'.[30] David Mure recounts Clarke's career in *Master of Deception,* and Clarke wrote in his own book *Seven Assignments* of his early war exploits (he was one of the fathers of the Commandos). But his plans to describe his deception task were never published: unlike Sir John Masterman, he was willing to conform to the security rules rather than flout them for his personal gain.

The beginning of the Second World War found Britain outnumbered and overstretched in every theatre, not least the Middle East, where the Italians menaced the British in Egypt from Libya and Ethiopia. In each of these countries the garrisons amounted to a quarter of a million men, while Wavell disposed of no more than 50,000 British and Imperial troops throughout the entire region, already charged with a difficult enough policing operation. Fortunately for the British, behind a taciturn and supremely undemonstrative nature Wavell was gifted with one of the most fertile and imaginative minds ever possessed by a British officer. He had made something of a reputation before the war writing pamphlets and giving lectures, and also produced an account of Edmund Allenby's campaigns in Palestine and a biography of his hero (on whose staff he had served), in which he outlined many of the methods that would later be used at El Alamein and in other places. 'Every commander', wrote Wavell in *Ruses and Stratagems in War,* 'should constantly be considering methods of misleading his opponent, of playing upon his fears, and of disturbing his mental balance.'[31]

The Italians invaded Egypt on 13 September 1940 and occupied Sidi Barrani, some sixty miles across the frontier, before settling down into a series of fortified camps. Wavell was looking immediately for ways to attack and in due course was in a position to launch Operation

COMPASS, originally intended as a 'five-day raid'. Various deception measures were put in place, mainly displays under auspices of the GHQ Camouflage Section, but their significance to subsequent events is obscure. As Richard O'Connor's Western Desert Force went on to sweep the Italians out of Egypt and across Cyrenaica, what was significant was that Wavell became convinced of the need for a dedicated unit to co-ordinate and implement tactical deception. Wavell wrote to London on 13 November 1940 requesting that Clarke (who had served on Wavell's staff in Palestine before the war) be sent to him to form a deception unit. Clarke arrived in the Middle East on 18 December.

Clarke started work with just two other officers and ten men in a converted bathroom in the GHQ building, and on 28 March 1941 the unit was officially designated Advanced Headquarters, 'A' Force. 'A' Force was a notional brigade of the Special Air Service, also only a notional body at this time, which supposedly existed in Trans-Jordan, from where it could intervene anywhere in the Middle East.* Eventually, this became 1st Special Air Service Brigade and Clarke's unit simply 'A' Force.[32] 'A' Force was always a small organization: after three years it amounted to just forty-one officers, seventy-six non-commissioned officers and three units of company strength. It did, however, employ numerous double agents

---

* The word 'notional' was first used in a deception context by Dudley Clarke to indicate something or someone imaginary to the deceiver but factual to the enemy.

in the various territories under British control – the Mediterranean, Iraq, Persia and much of Africa – most notably the CHEESE network. It later comprised a small mobile headquarters operating through an 'Advanced HQ West' for the Allied Commander-in-Chief, a 'Tactical HQ West' for 15th Army Group, and an 'Advanced HQ East' for Iraq, Iran and East Africa. There was also a 'Tactical HQ East' to serve with any commander operating independently in the Middle East.

'A' Force worked very closely with the GHQ Camouflage Section and used a great deal of specialized kit, much of it manufactured themselves: the standard-issue overhead artillery camouflage net, for instance, designed for use in temperate climes, was found actually to make guns more conspicuous in the desert. Wherever possible, local materials were used.[33] 'A' Force began with Dudley Clarke (then a lieutenant-colonel, Royal Artillery), Captain Victor Jones, (14th/20th King's Hussars and an expert in visual deception) and the small MI9 organization with which it was amalgamated. MI9 was concerned with the recovery of escaped Allied prisoners of war and others through Arab agents, who were paid a price per head for men they brought into British lines. When 'A' Force grew sufficiently to require headquarters of its own, these were established above a bordello at 6 Kasr el Nil in Cairo.

According to the MI9 historian M. R. D. Foot, 'A' Force was combined with MI9 for purposes of cover, since deception was more 'secret' than escape. But the two units could help each other and there were a number of personnel in Lieutenant-Colonel Tony Simmonds' MI9 staff who were regarded as interchangeable. Most notable was Jasper Maskelyne, who worked for MI9 but describes himself in his own book, *Magic: Top Secret* as the inventor of the dummy tanks, submarines, aeroplanes and landing craft that were the essential ingredients of physical deception. Doubtless his advice was sought and fully acted upon, but this work was principally carried out by GHQ Camouflage Section.[34] At the same time, in Greece, Crete and Syria MI9 operations enabled 'A' Force to spread its influence throughout the Eastern Mediterranean.

With the defeat of the British in Greece and Crete, and their expulsion by Generalleutnant Erwin Rommel from all of Cyrenaica except the port of Tobruk, things looked bleak by the middle of 1941 and Wavell swapped places with the Commander-in-Chief India, General Sir Claude Auchinleck. Auchinleck faced threats to Cyprus, which had only a 4,000 man garrison, and it was now that operational deception began. Clarke had always made a point of building up a false and exaggerated order of battle. As he pointed out, too few British generals were blessed with adequate reserves and the best way to fill the gap was by the orderly, consistent and methodical building up of the false order of battle. The enemy appreciation of British (and later American) forces was increased by the addition of notional brigades and divisions, and later even corps and armies. The main method of giving these forces the appearance of reality was by foisting their identity on various non-combatant and even static formations and organizations: areas, sub-areas, training schools, depots and other base facilities. These were renamed as divisions, brigades and units and their vehicles wore divisional signs designed by 'A' Force. This was helped by the fact that the British and Americans wasted far more manpower than the Germans on back-up and purely administrative duties. Sufficient radio traffic was generated (if not naturally, then by simulation) so that once a bogus formation had been placed and identified (usually by double agents), it behaved on the air like a real one.

The build-up of this notional order of battle, especially in the early days when the Middle East was swarming with spies and informers (nearly 400 were apprehended in Syria and Lebanon alone), was 'a dull hard slogging business'. In due course it would nevertheless repeatedly prove its value. It had no glamour attached to it, but more exciting methods such as the running of double agents would have been toothless without it. It also meant deceiving the majority of one's own side and indoctrinating preoccupied and incredulous officers into a procedure that must have seemed bizarre. Sometimes the dummy formations were provided with dummy equipment, and finally, as the role of the double agents became firmly established, the notional order of battle became the substance of their reports. When Wavell asked Clarke what 'A' Force was worth to him, Clarke was able to reply 'three divisions, one armoured brigade and two squadrons of aircraft'.

Eventually three bogus armies were built up and these were to loom large in the minds of the German High Command, supporting their misconception that the British were planning to land in the Balkans in order to seal off central Europe including Germany from the

Russians sweeping in from the east. The identities of these false formations were leaked by every means available. The system could be tiresome, particularly to the administrative staffs, but this was a small price to pay. By the end of 1943 the twenty extra divisions in the three notional armies were so firmly fixed in the minds of the Germans that they were far more menacing than any build-up of British and American forces in England.[35]

In the meantime, however, Rommel was in no way deflected by the notional 10th Armoured Division, created by Jones out of dummies. But the '7th Division' was created to 'reinforce' Cyprus with a full programme of visual displays involving camps being erected and movement and administrative orders being issued in abundance. Cyprus became a showpiece of deception. The garrison was commanded by a 'lieutenant-general' (a substantive brigadier), the landing strips were adorned with 'Spitfires' and 'Hurricanes' (four out of five of which were canvas and wood) and the Cyprus Regiment's transport was decorated with divisional and corps signs.[36] A complete 'defence plan' was 'lost' in Cairo and when the Axis were again expelled from Cyrenaica during the CRUSADER battles of November and December 1941, captured intelligence summaries revealed this yarn had been completely accepted.[37]

For Operation CRUSADER in November 1941 a number of operational deceptions were attempted. Fake concentrations of troops were assembled at the Siwa and Giarabub oases far to the south (captured the previous year with the assistance of dummy parachutists who proved so successful that the Italian garrison fled before any real British troops arrived). Camps were laid out, complete with the cookhouses, latrines and slit trenches necessary for a force of divisional size. Following the battle, however, an inquiry into the deception reported that there had not been enough vehicles or materials for fake tentage to give the false concentrations a genuine feel, and more critically, synchronization with the genuine operation had been poor. Crucially, the lesson was learned that deception was unlikely to succeed unless planned with the same thoroughness as a genuine operation. When captured documents later enabled the British to see the results from the enemy perspective, they did find the Germans estimated the force at Siwa and Giarabub to consist of an infantry brigade, two or three armoured car units and the Egyptian Camel Corps (although no effort had been made to simulate camels), but the vital question of whether the enemy altered his dispositions in any way as a result of the deception remained unanswered.

More successful were efforts to protect the vulnerable railhead at Capuzzo, which was crucial to supplying Eighth Army. For this a dummy railhead was built at Misheifa, ten miles closer to the front, which was also positioned so as to help the enemy draw the wrong conclusions regarding British intentions. Having overcome a shortage of rails by making dummies from 'flimsies' (four-gallon tins used for carrying petrol and water), the line was laid out at the speed a genuine railway could be laid and a fake train built with wood, string and canvas. During the build-up to CRUSADER this 'railhead' was bombed eight times while the genuine railhead went unscathed. A map found later on a crashed German aircraft showed Misheifa as the rail terminus.[38]

As 'A' Force expanded, a special depot was established in Cairo for the production of increasingly large amounts of dummy equipment: a complete brigade was formed – 74th Armoured Brigade – that was not so much notional as chimerical. The new commander of Eighth Army, Major-General Sir Alan Cunningham, wrote to GHQ that they could simulate feints on enemy flanks, mix dummy units with real ones to exaggerate their size, and 'enable real tank units to move and be rapidly substituted by dummy units thereby misleading the enemy to our real strength and dispositions'. Apparatus was devised for creating and concealing tank tracks and a unit formed to operate them, 101st Bn, Royal Tank Regiment. Progress was also being made with sonic deception, recording the noise of tank movements and broadcasting them through amplifiers provided by a political warfare (Psyops) unit.[39]

Unfortunately, the proliferation of interest in deception, and of personnel involved, caused serious problems as various factions tried to exercise control. In October 1941 Cunningham's chief of staff, Brigadier J. F. M. (later Sir John) Whiteley, tried to resolve it by separating 'strategic' deception, to be controlled by 'A' Force, and tactical and operational deception in the field, which he recommended should be run by a staff officer of Eighth Army, to handle 'planning and development of deception units and schemes for the control of camouflage'. Special deception officers were to be attached to corps, division and brigade headquarters.[40] While this appeared fine on paper, it made Clarke very unhappy. He had learned in 1941 that what differentiated deception in the Middle East from its counterpart in London was its centralized control: a single hand drew the plan, which would be implemented by staff officers or failing them by regimental ones, trained to obey orders without adding refining touches of their own [41] Besides, there were no facilities for training 'deception officers' and consequently CRUSADER went ahead with no tactical deception at all. Auchinleck became aware of the shortcomings in the deception organization only in February 1942. He immediately ordered that all deception should be the responsibility of 'A' Force, which would answer directly to the operations branch of GHQ.

Almost immediately Clarke began laying the basis of later strategic deception in the Middle East. In March 'A' Force commenced Operation CASCADE, the first comprehensive order-of-battle deception plan covering the whole Mediterranean theatre. This would in due course also give Clarke the flexibility to implement the operational-level deception plan that would support the Battle of El Alamein. During the remainder of 1942 'A' Force created a bogus armoured division (15th) and seven bogus infantry divisions (including two from India and one from New Zealand) as well as a bogus corps (XXV) to add to the 10th Armoured and 7th Divisions.

However, nothing was more important to all deceivers than the breaking of the Abwehr Enigma ciphers at Bletchley Park in December 1941. Now ULTRA extended into the enemy's thoughts on British intentions, enabling deception to be altered accordingly, and to tailor plans to fit German means of information gathering.[42] Moreover, the chances of success were greatly increased since it revealed German opinions on deception operations, enabling their effectiveness to be monitored. It also highlighted the rivalry that existed between the plethora of intelligence agencies working in Germany, a result of Hitler's policy of denying too much power to any one subordinate or organization. Increasingly, 'A' Force's deception operations could take on a strategic dimension. But in the meantime there was still the small matter of beating Rommel, who during the summer of 1942 drove Eighth Army back towards the Nile delta, and was only finally halted near the otherwise insignificant railway station of El Alamein. Auchinleck was replaced by General Sir Harold Alexander, and Montgomery was appointed commander of Eighth Army.

Meanwhile it had become clear that a tightening of security in Egypt was necessary. The Egyptian court was the first target, as King Faroukh came under British pressure and the weak premier, Sirry, was replaced with the British nominee, Nahas Pasha. At the same time the brood of anti-British courtiers were cleared out and the German CONDOR spy ring closed down. This was causing consternation at the time since it appeared that Rommel was receiving high-grade intelligence from within the heart of GHQ. The ring was traced, largely thanks to ULTRA, to a houseboat on the Nile, but it subsequently transpired that this was not Rommel's source. On 10 July 1942 at Tel el Eisa the Australians overran 621 Radio Intercept

Company, commanded by Hauptmann Alfred Seeböhm, who was mortally wounded. The company's records were found to reveal not only the dreadful lack of proper voice procedure used by British radio operators but also that Rommel's high-grade intelligence was coming from the American military attaché, Colonel Bonner Fellers, whose 'black code' had been broken and whose reports to Washington, detailing every last item of minute interest, were being read by the Axis. Although a replacement company was sent from Germany, it never achieved the efficiency of Seeböhm's unit, and Rommel found himself relying increasingly on agents' reports, which meant CHEESE. The Germans were thoroughly impressed with what they had received so far and Rommel requested that CHEESE's reports should be available in his caravan early each morning.[43]

Soon after Montgomery arrived in theatre, he summoned Clarke to see him on 19 August. 'A' Force was at the time trying to divert reinforcements from Rommel by mounting a notional threat to Crete from Cyprus (Operation RAYON), apparently with little success. While efforts to create the false order of battle were later shown to have been successful, they were as yet having no discernible effect on the German High Command. Rommel attacked as soon as he had sufficient fuel but the British defensive position previously selected by Auchinleck at Alam Haifa was too strong to turn. In the aftermath of the Battle of Alam Haifa, Montgomery was criticized for not having launched an immediate armoured counter-attack to push the Germans off Himeimat Hill, which gave a commanding view of the southern sector of the Allied line. Apart from the fact that it would surely have ended with a repetition of British armoured formations dashing themselves against the rock of German anti-tank defences, with his deception plan already in mind Montgomery asked what would be the point of constructing dummies 'if the Germans cannot see them? Leave them in possession of Himeimat. That is where I want them to be.'[44]

BERTRAM was to be the largest operational deception plan of the war thus far, and a comprehensive radio programme was devised to support the visual illusion that identifiable, signature formations for any forthcoming attack were in fact in the south.[45] But, as Clarke had long since discovered, operational deception could not be separated from the strategic deception that he was developing concurrently. As composer of the piece, he would be away for this first combined showing of its capabilities, persuading the sceptical Americans of its value. The conductor responsible for the detailed running of the plan was his deputy, Lieutenant-Colonel Noël Wild, who joined 'A' Force in mid-1942, just as the disastrous Gazala battle was beginning. A plan to suggest that Montgomery did not intend to attack at all was implemented through the double agent network (Operation TREATMENT). This had been enhanced by the autumn of 1942 by the double agents PESSIMIST and QUICKSILVER, whose credibility was established by a string of easily verifiable truths. PESSIMIST now reported a troop exodus across the Syrian desert. (It should be remembered that the Battle of Stalingrad was beginning and it seemed little could prevent the German drive across the USSR from reaching the Caucasus and its oilfields.) The success of the combined plans meant that when the great battle opened on 23 October Rommel was absent in Europe (on sick leave). Only when he returned to command twenty-four hours later did he regard the situation as 'serious'.[46]

The most important aspect of BERTRAM was that here were all the ingredients of future deception operations working together. Although essentially an operational-level plan, it included significant strategic elements such as the false order of battle and the use of double agents' reports, together with false radio traffic and concentrations of dummy landing craft, all backed up by detailed and co-ordinated tactical displays. It is also notable that the conductor of this scheme, Noël Wild, went on to run the deception for the invasion of Normandy.

CHAPTER SIX

STRATEGIC DECEPTION

'There is required for the composition of a great commander ... an element of legerdemain, an original and sinister touch which leaves the enemy puzzled as well as beaten.'

Winston Churchill

SRAEL TOOK THE world by surprise when it launched a preemptive attack against the Arab states in 1967, but the Arab nations turned the tables in spectacular style six years later. Any ideas that surprise could no longer be achieved in a world of spy satellites and high-technology surveillance equipment must surely have been dispelled by the time Egyptian engineers were breaking down the defences along the eastern side of the Suez Canal, the product of clever military deception and the 'noise' prevalent in Arab politics and culture (demonstrated again in August 1990, when Iraq invaded Kuwait). At the strategic level, 'noise' becomes a crucial factor in the success of deception. It exists in various forms – cultural, political and military – and can be enhanced or created by the deceiver. Western analysts, for example, have difficulty predicting the course that Arab states may adopt since in Arabic culture verbal articulation of imagery is as important as reality:

> Arab speech ... tends to express ideal thoughts, and to represent what is desired or hoped for as if it were an actual fact. There is thus among the Arabs a relatively greater discrepancy between thought and speech on the one hand and action on the other.[1]

Thus many political acts are made for their symbolic value, which leads to difficulties of interpretation.

President Abdel Nasser's posturing before the Six-Day War of 1967 may have led the Israelis to conclude that similar Egyptian belligerence before the Yom Kippur War of 1973 was nothing but bluff.[2] The Six-Day War was the product of spiralling tensions which escalated beyond control. Syria, in the throes of internal dissent, externalized the threat towards Israel, and Israel countered with its own threats. In the middle of May Nasser mobilized Egyptian forces and deployed along the Israeli frontier in Sinai, a repetition of a similar situation in 1960. The Israelis, who had until now considered the various manoeuvres as a bluff, felt genuinely threatened and decided to act. They launched a devastating attack, in which their air force came in from the sea to catch their Egyptian counterparts on the ground and destroy them. It seems that all Nasser's actions were taken in full awareness that Egyptian forces were no match for the Israelis, and that he was acting out of a need to score political points as the self-appointed leader of the Arab world.[3] However, Israeli analysts detected a similar pattern when Anwar Sadat prepared to launch the Yom Kippur War six years later, only to find that this time Egypt was not bluffing.[4]

The effect of 'noise' is amply demonstrated by the imposition of a curfew on Baghdad on 28 September 1973, in reaction to domestic problems and the apparent threat of a *coup d'état*. While in no way connected with Egyptian and Syrian deception plans, it coincided nicely with them and aided them. Similarly, a Palestinian group calling itself the 'Eagles of the Palestine Revolution' hijacked a train in Austria bound for a Jewish transit camp in Schönau. Many analysts believe this was part of Syrian efforts to divert attention away from the Middle East and it certainly provided the Syrians with an excuse for a military buildup against possible Israeli retaliation.[5]

The Arab media contribute considerably to the 'noise', because while being very tightly controlled by the various governments and thus engaged in disinformation, they are also capable of producing extremely accurate reports, usually for political purposes. They are also prone to exaggeration. Four days before the Yom Kippur War the Middle East News Agency (MENA) reported the increased state of alert in the Egyptian Second and Third Armies. The significance of this story was obscured, however, by a series of reports from all over the Arab world of war-scare stories. These were partly a result of the MENA story itself and of Israeli responses to it, but it was all too easy for the central story to be drowned out by the resulting hubbub.[6] The Israeli prime minister, Golda Meir, thought so:

> I have a terrible feeling that this has all happened before, it reminds me of 1967, when we were accused of massing troops against Syria, which is exactly what the Arab press is saying now. And I think it all means something.[7]

It was within this context that Egypt and Syria made genuine preparations for war.

The Japanese attack on Pearl Harbor on 7 December 1941 is possibly the most famous act of strategic surprise, and excites controversy to this day. While some writers remain convinced by a conspiracy theory view of events, the reality is much less dramatic: the attack was a simple failure of intelligence collation. Accusations that President Franklin D. Roosevelt deliberately allowed the attack to take place are not supported by evidence and any suggestion as to motive is very tenuous.[8] The Japanese and the Americans both made mistakes as a result of serious misunderstandings of their opponents' mentality and intentions.[9] By November 1941 war was already seen as inevitable, as indeed it had become with the American imposition of an oil embargo in July. But crucially, the war was regarded as being most likely to engulf the western Pacific (as indeed it did).

Negotiations dragged on between the two sides throughout the autumn but the Americans had access to Japanese high-grade diplomatic signals (code-named

MAGIC), and it was not a lack of information but a failure adequately to collate the indications of forthcoming hostilities, together with failures to disseminate what intelligence had been processed, that led to the Japanese achieving overwhelming surprise on that fateful morning. Although there was an abundance of indicators, there was no single intelligence organization to see the overall picture. Richard Helms (later head of the CIA for seven years) stated that the CIA 'was felt necessary because the Japanese attack on Pearl Harbor in 1941 clearly demonstrated that if the relevant intelligence available in Washington had been fitted together in a central place, and examined, we could have foreseen the attack'.[10] Added to this high-level failure, there was a fundamental split between the two services (there was as yet no independent air force) and further splits within them.

When the British sent their double agent TRICYCLE (a Yugoslav called Dusan Popov) to the USA in August 1941, he took with him an example of the newly invented microdot – given him by his German controller – which included a highly detailed questionnaire on the defences of Pearl Harbor. In Washington he met J. Edgar Hoover, Chief of the Federal Bureau of Investigation (FBI), which at that time was responsible for counter-espionage. Hoover dismissed the *louche* playboy (whose code-name supposedly derived from his preference for two women at a time) out of hand. He not only overlooked the questions in the microdot but also scuppered the whole security liaison effort with the British. Hoover failed to inform the Navy or State Departments about the questionnaire (of interest to the Japanese as much as the Germans now that the Axis was pooling its intelligence gathering) and glibly boasted in his memoirs of interviewing a 'dirty Nazi spy' and 'sending him packing'. Not only did Hoover fail to understand the importance of TRICYCLE, his mission and what he brought with him; he did not care.[11] In fact, the Japanese war machine remained geared towards South-East Asia rather than the central Pacific and remained ready until the last minute to cancel the Pearl Harbor operation if necessary. They did institute operational deception measures – radio transmissions were made in the Inland Sea and at Kyushu while the fleet remained on radio silence – but they relied more on tight security than active strategic deception to gain the advantage.[12] The 'noise' that concealed the Japanese moves was largely generated within Washington itself.

OPERATION BARBAROSSA

Strategic deception had proved essential six months previously, when the world's bloodiest and most brutal war was launched with the Nazi invasion of the USSR on 22 June 1941. It has long been a source of amazement that this colossal assault could have been achieved with complete surprise. Yet by a combination of active and passive deception Nazi Germany mounted an attack with 150 divisions against the largest state in the world. In this respect, according to Barton Whaley, Hitler succeeded in making Stalin 'quite certain, very decisive, and *wrong*'.[13] Again there were ample indications of what was coming, but in this case considerable 'noise' was generated by the Germans and amplified by the mind of the one man that mattered: Josef Stalin. At the lower levels German deception of their own troops and the civilian population through propaganda, troop orders and diplomatic channels contributed to operational security. The considerable evidence that was available was either overlooked or deliberately ignored or rejected by Stalin, proof that the focus of the deceiver must be directly on the decision maker.

All executive power was concentrated in Stalin's hands, and he consistently refused to accept the hypothesis that the Germans would attack during 1941, since he was convinced that they could not attack before mid-1942.[14] Both Andrei Gromyko and Nikita Khrushchev later noted that Stalin was convinced Hitler would keep his word, partly because he was petrified of a German attack and partly because once he adopted an idea, nobody was able to change his mind.[15] He believed that any German build-up would follow the previous pattern of demands and provocation, which the Soviets would be able to recognize and parry in 1942, and that the Germans would not launch an attack while still continuing a second (albeit largely air and naval) campaign in the West. So convinced was Stalin that the Germans would keep their side of the bargain that officers who suggested otherwise were liable to arrest as *provocateurs*.[16]

Hitler's first directions to his general staff to prepare plans for an invasion of the USSR (Operation BARBAROSSA) were given as early as July 1940, before the Battle of Britain had even begun. He had, in fact, compelling reasons for attacking the USSR in the summer of 1941: he was certainly convinced of the necessity and, given that he possessed no land frontier with Britain and given how close he came to success in that first year, the theory that it was necessarily fatal to start a 'war on two fronts' is unsustainable. In fact, Hitler used operations conducted in one war to create the deception necessary to have a chance of success in the second.[17] Although he hoped that a successful air campaign and, if necessary, invasion (Operation SEALION) would bring Britain to defeat, a reading of *Mein Kampf* shows his real concern had always been with the hated Bolsheviks and the Slavic peoples, whom he considered subhuman. It was in the East that his true dreams of finding *Lebensraum* for his mighty new Reich were centred.

The USSR possessed the largest army in the world at the time and Hitler believed it was getting stronger, in spite of Stalin's purge of the officer corps from 1937 and the setbacks inflicted during the winter war with Finland. With the defeat of the Luftwaffe and the onset of winter, Hitler quietly ordered the indefinite postponement of invasion plans in the West and on 18 November he signed Directive No. 21, ordering preparations for the invasion of the USSR to be completed by 15 May 1941.[18] SEALION was kept outwardly in being to maintain pressure on Britain and to distract attention from Hitler's purposes and preparations in the East.[19]

In a masterful double bluff troop movements intended for BARBAROSSA were to be 'seen as the greatest deception in the history of war', supposedly a mask for the final phases of the invasion of Britain. By October 1940 troop concentrations in the East had already jumped from five to thirty-three divisions, including five panzer, two motorized and one cavalry division, together with training, logistics and communications facilities and dozens of new airstrips. Concealment was no longer possible, so the German military attaché in Moscow informed the Soviets that the older men previously stationed in the East were being replaced by younger men, to free the former for war production and because the training facilities were both better in the East and free from British air attack. There was 'no reason for the [Soviets] to be alarmed by these measures'. When Stalin wrote to Hitler in early 1941 to say he was aware of the build-up and had the impression Hitler was about to attack, Hitler wrote in reply, confident that nobody beyond Stalin's immediate circle would see his letter, that the troop concentrations served an entirely different purpose: to regroup and retrain free from British interference. Meanwhile German troops in the East were given the impression they were taking part in a deception operation and to provide 'rear cover' for an invasion of Britain.[20]

It is tempting to think that the signing of Directive No. 21 rendered all other operations subordinate to Barbarossa, but two grand deceptions surrounding the operation comprised events and circumstances that existed before the decision to launch it was made. The Nazi-Soviet Non-Aggression Pact of August 1939, which had permitted Hitler to begin his campaign against Poland and the subsequent invasion of the West, and the continuing resistance of Britain both suggested the the broad strategic outlook remained unchanged. It was these ongoing historical events that the Germans consciously used as deceptive elements once the final decision to invade was taken.[21] In Directive No. 23 Sealion was maintained with orders to inflict maximum possible losses on the British and the result was the *Blitz* (the bomber offensive aimed at major cities) of September 1940-May 1941, 'to give the impression that an attack on the British Isles [was] planned for [1941]'.[22]

Two further deceptions were initiated, under the code-names HARPOON and SHARK, intended to convince the British that an invasion was planned for around 1 August 1941. SHARK, which comprised preparations by troops in France and Scandinavia, was ordered in April and HARPOON, which took the form of naval operations, in May of that year. During the winter and spring various other operations directed against Britain served to reinforce the general deception regarding BARBAROSSA. Hitler's negotiations with the Spanish dictator, Francisco Franco, helped to focus attention on the western Mediterranean, as did deteriorating relations between the Germans and Vichy France. Neither of these situations was intended as a deception, but both were convenient.[23] Meanwhile, with the approval of the pro-Axis government of General Ion Antonescu, Hitler sent German forces into Romania on 8 October 1940, a move that served to cloud relations with the Soviets as it raised the question of spheres of influence within the Balkans. Hitler was forced to secure the Balkans against British interference in BARBAROSSA, but at the same time genuine operations mounted against the British in the eastern Mediterranean and their control of the Suez Canal could be plausibly incorporated into the general deception of continuing war against the recalcitrant islanders.

A stable situation in the Balkans was the minimum requirement for the successful prosecution of BARBAROSSA. A stalemate between the Italians in Albania and the Greeks, whose country Benito Mussolini had decided to 'occupy' in a fit of pique at not having been informed in advance of Hitler's occupation of Romania, was regarded as acceptable since it was unlikely to lead to direct British intervention. Ironically, Churchill was extremely keen to intervene, but the Greeks refused his offer since the aid available from British forces in Egypt was too little to be effective and might provoke the Germans. Although some elements of BARBAROSSA were due to emanate from Romania, most were concentrated north of the Carpathians.[24] However, the overthrow of the Yugoslav government following its accession to the Axis, and its replacement by a neutralist government under General Dushan Simovich and the newly proclaimed King Peter II on 27 March 1941, forced Hitler's hand. He immediately ordered the invasion of Yugoslavia in order to safeguard the southern flank of BARBAROSSA, a move that also provided deceptive potential since it supplied an excuse for the movement of hundreds of thousands of troops into Eastern Europe. Two subsequent German operations, MARITA (directed at Greece in April, after the Greeks had finally accepted British aid) and MERCURY (against Crete in May), also served to suggest a reduced possibility of a German attack on the USSR during the summer of 1941.

Therefore, German intervention in the Balkans, designed to secure the south-eastern flank of BARBAROSSA, also provided a plausible deceptive opportunity. The specific deception was one of a misleading nature which, despite a little ambiguity, helped explain the build-up of troops in Poland. The presence of 60,000 British and Commonwealth troops in Greece helped refocus attention on the war with Britain and helped the Germans to mislead the Soviets about troop movements in Hungary, Poland and Romania, which could be passed off as part of the Balkan operations.* By the same token the transfer of German forces, including the Deutches Afrikakorps, to assist the Italians in their campaign against the British in North Africa, followed by spectacular Afrikakorps successes in April and May, further focused attention away from the quiet build-up taking place in Poland.[25]

Not that indications of the forthcoming invasion were unavailable. Gunnar Hagglof, responsible for German affairs in the Swedish Foreign Ministry, described the tearful reaction of the Soviet minister in Stockholm,

---

* It is also worth noting that the Balkans campaign was not responsible for a fatal four- to six-week delay that led in due course to the German invasion becoming bogged down in the Russian winter. The expansion of the German Army and the associated demand for stores led OKW to realize that German production capacity was insufficient, rendering changes to the timetable necessary (M. van Creveld, 'The German Attack on the USSR: The Destruction of a Legend', *European Studies Review,* vol. 2, no. 1 (January 1972), p.83).

Alexandra Kollontai, to his indications that Germany was preparing an invasion. He had no right to tell her she said, and she had no right to listen. Orders had obviously been received from the highest authority that such rumours were to be ignored or denied.[26] The situation being created could not be allowed to interfere with the normal smooth running of Soviet-German relations and Moscow now commenced on its own disinformation campaign, mainly through the German Consul in Harbin, August Ponschab, which was designed to reassure Berlin that the Soviets posed no threat. As became painfully obvious on 22 June 1941, this was wholly ineffective but its pursuit served to make German deception ultimately more effective. In March, when Admiral Nikolai Kuznetsov gave instructions to shoot down German aircraft that were clearly involved in photographic reconnaissance missions, he was peremptorily ordered by an angry Stalin to withdraw the instructions. Nothing was to be allowed to interfere with a continuing accommodation between the two dictators.[27]

Yet rumours of the intended invasion had been reported to Moscow within days of Hitler's Directive No. 21 being issued, and in February 1941 Moscow received information including the outline plan and a tentative date for commencement of 20 May. The Glavnoye Razvedyvatelnoye Upravlenie (GRU, the Soviet military intelligence) had considerable evidence of an impending attack, but it also believed that the Germans would await a victorious conclusion in the West, a not unreasonable premise.[28] Meanwhile, the Abwehr stepped up its spying activities in the USSR, albeit with little success (their failure accounted in large part for Germans' appalling lack of quality intelligence on Soviet capabilities), but none of this aroused particular suspicion in Moscow.[29]

The second major deceptive element, the Nazi-Soviet Non- Aggression Pact, served to reinforce Stalin's convictions. It was a totally pragmatic agreement between two ideologies diametrically opposed to one another. It nevertheless appeared to give both sides what they wanted, and both made considerable territorial gains as a result.* Throughout its existence Stalin made sure the Soviets kept their side of the bargain to the letter, sending trainloads of food and raw materials westwards right up until the day war began. Whenever strains appeared, such as those following the German move into Romania and the subsequent operations in the Balkans, the Nazis were always able to pass them off as being directed at the British. High-level diplomatic contacts and further trade agreements were concluded and the state-controlled media of both countries poured forth soothing propaganda for both internal and external consumption, contributing enormously to the surprise of both populations when the cataclysm came.[30] Deception of the German public (and indirectly the Soviet government) was carried out in the form of radio and press releases, the dissemination of rumours, orders to troops and preparations for state receptions. On

---

* Between 17 September 1939 and 26 June 1940 the Soviets either annexed or militarily occupied eastern Poland, parts of Finland, the previously independent Baltic states of Lithuania, Latvia and Estonia, and the Romanian provinces of Northern Bukovina and Bessarabia, a total of approximately 180,000 square miles.

13 June 1941 Hitler's propaganda minister, Josef Goebbels, wrote in the Nazi organ *Völkischer Beobachter* that the fall of Crete signalled the imminent invasion of Britain. In a clever ploy the entire issue was immediately confiscated by the police as soon as it reached the foreign press corps.

On the radio, musical programmes in which German soldiers sent in requests were used to suggest that first-class formations were stationed in the West and only defensive and training formations were stationed in the East; for example: 'Members of the Leibstandarte [an SS formation originally formed from Hitler's bodyguard] send their wounded company commander three bottles of Hennessy and wish him a quick recovery.'[31] The reference to French brandy was designed to suggest that these troops were somewhere in the West. As the build-up continued in Poland during the spring of 1941, it became increasingly difficult to conceal the truth, but it was put about that these were deceptive measures designed to put the Soviets off during the attack on Britain.

Although the British had ample evidence of German troop movements, these did not seem necessarily to entail the invasion of the USSR. Such a possibility seemed 'too good to be true' to Churchill, until he saw the originals of pertinent intelligence reports. It was the odd movement of five panzer divisions that crossed Romania on their way to Yugoslavia and Greece, having been routed to Krakow before the Yugoslav *coup d'état,* that 'illuminated the whole European scene like a lightning flash'.[32] Once convinced by his own intuitive insight, Churchill assembled the evidence and tried to present it to Stalin. But Stalin was convinced that all efforts by Churchill to demonstrate the imminent peril were merely attempts to provoke the Soviets into counter-measures against the Nazis and to disrupt their (so far) peaceful relations. The British passed a warning on 19 April. The reception it received can be gauged by that of another. On the same day Ismail Akhmedov, the Tartar acting chief of the GRU's fourth division (technical espionage), received a report from a source in the Czech Škoda works that provided convincing evidence of a planned attack on the USSR between mid and late June. On it was scrawled in red pen: *Angleyskaia Provoatsiia Rassledovat! Stalin* ('English provocation investigate! Stalin').[33]

The flight of Hitler's deputy, Rudolf Hess, to Scotland on 10 May 1941 in an abortive attempt to secure peace surprised everyone; it also served to heighten Stalin's deep-rooted mistrust of the British.[34] One possible reason for Hess's flight (which he claimed to have attempted twice before) was his rabid anti-Communism, combined with his ignorance of the real purpose of the trade negotiations that the Germans were then pursuing with the Soviets. These were actually part of their continuing deception, necessary to finalize their plans and preparations, which had been delayed. The build-up in Poland could also be passed off as no more than an attempt to 'squeeze' the Soviets in the negotiations. The Soviets also used the talks as a deception operation, hoping to satiate the Germans and defuse any pretext for an attack. They were prepared to prolong and extend them until it was too late in the year for the Germans to contemplate invasion, during which time they could continue their rearmament programme.[35]

As May progressed, General Georgi Zhukov correctly suggested that the Germans were capable of launching a surprise attack and proposed that the Soviets launch a pre-emptive attack of their own. Stalin quickly ruled out such an option, but the warnings increased. The prize agent in Japan of the NKVD (Narodny Kommissariat Vnutrennihk Del, one of Stalin's state security agencies), Richard Sorge – who had repeatedly sent warnings which Stalin contemptuously dismissed – sent another four in May, the last of which stated the attack would begin without the benefit of an ultimatum or declaration of war. Stalin dismissed this too; he called Sorge 'a little shit who has just set himself up with some good business in Japan' and Sorge was reduced to weeping 'Moscow doesn't believe me' in his mistress's arms.[36] As May turned into June the Kremlin rejected warnings from other sources, including the German ambassador himself, the anti-Nazi Graf Friedrich von der Schulenburg, and his counsellor of mission, Gustav Hilger. Their approach to Vladimir Dekanozov, the Soviet plenipotentiary at the trade negotiations, was rejected out of hand by Dekanozov, who had 'no comprehension of the good will' that motivated them.[37] Nevertheless, despite the political culture surrounding Stalin, the substance of the conversation was forwarded to him. It was apparent that Stalin was convinced that Hitler was 'bluffing'. In conversation with his toadies in 1942 Hitler took credit for 'making the Russians hold off right up to the moment we launched our attack ... by entering into agreements which were favourable to our interests'.[38]

The stakes in these negotiations were soon raised beyond anything a reasonable government, let alone Stalin's, could accept. The purpose was to prolong the talks into early June.[39] By now even the British intelligence authorities who had not shared Churchill's conviction from the outset were convinced of German preparedness to invade Russia. ULTRA decrypts showed a steady build-up of Luftwaffe units in Poland and the Joint Intelligence Committee (JIC) forecast the launch of the operation as being between 20 and 25 June.[40] By the beginning of June east-bound rail traffic was so extensive that it 'became obvious even to the layman that large-scale troop concentrations were taking place'.[41] When Zhukov and Marshal Semyon Timoshenko, the Defence Minister, went to present this evidence to Stalin, the latter countered with his own documentation, which was remarkably similar in all but one crucial aspect. The GRU chief, General-Lieutenant Filipp Golikov, knew Stalin's view that there would be no war for another year and had supplied him with refutation of Timoshenko's information, in particular of Sorge and his prediction that an attack would be launched on 22 June. Golikov classified as reliable all reports confirming that German deployments were part of SEALION.[42]

Meanwhile, the German build-up continued. The deployment in the East of many Luftwaffe formations was particularly difficult to justify in terms of a defensive posture. However, during the campaign in the West, in order to conceal preparations for that campaign by not deploying high-level headquarters, the Luftwaffe relied heavily on Luftflotte 1 (1st Air Fleet), based in Berlin and with base areas in East Prussia. The Soviets now accepted its presence as natural and reasonable and it was given the task of attacking the entire front on the commencement of the invasion. The task of collating target intelligence fell to Major Rudolf Loytved-Hardegg, who was forced to resort to many imaginative deceptions in order to gather his information. Dummy Lufthansa facilities were established in Helsinki, which operated flights to Moscow and other key points staffed entirely by Luftwaffe personnel. Two other dummy aviation businesses were set up in East Prussia: one an air-mapping service and the other a pilot training school, which systematically sought out Red Air Force targets along the border. German immigrants arriving from the USSR were carefully screened and from one such immigrant Loytved-Hardegg learned of the advanced nature of the Soviet aviation industry.[43]

Operational security involved retaining aircraft well to the west before 'B-Day', continuing the bomber offensive against Britain throughout the spring and, by flying multiple sorties, making the effort appear greater than was really the case.[44] Oberkommando der Luftwaffe (OKL, the Air Force High Command) took steps to make it appear that all of Luftflotten 2 and 3 were involved. Generalfeldmarschall Albert Kesselring, commanding Luftflotte 2, which was due to operate behind Army Group Centre in the invasion and which controlled around half the total air resources committed to it, spent as much time as possible visibly at his headquarters in Brussels rather than his alternative headquarters in Warsaw.[45]

The air forces operated from bases no closer than fifty kilometres from the Soviet border, but army formations could not start from so far back. The concentration of almost 150 divisions could not be completely concealed but to reduce its significance the border area itself was only very lightly held. The imposition of strict radio silence and the use of peacetime railway schedules helped to reduce the impact of these movements and by restricting information to the reason for the deployment security was maintained. The first mention of BARBAROSSA in a corps war diary did not occur until 15 April, and the divisional commanders and key staff were not informed until a month later.[46] In fact, the German troops were themselves deceived as to their presence in Poland, which seemed very difficult to explain. The reasons they were given included the better training facilities, security against Soviet moves during the 'invasion' of Britain, and significantly, that they were deploying to the east as a deception for SEALION, the story they also wished the Soviets to believe. Oberkommando des Heeres (OKH, the German Army High Command) issued orders for embarkation drills and plans to be issued for the transfer of the bulk of the troops back to the West. As late as May the troops were receiving English lessons and planning was being conducted on recently issued maps of Britain.

In the last few weeks the movements increased in intensity: twenty-eight panzer and motorized divisions moved from France and Germany, and eighty-six divisions already in the East moved closer to the border. The ninety-six infantry divisions of the first wave were marched up to the frontier on foot at night and concealed in the large forests during the day. The thirty-one panzer and motorized divisions of the first wave were moved only four days before the attack and echeloned further back from the actual frontier. By thus restricting contact with the population the Soviet agent networks were neutralized.[47] Only in the afternoon of 21 June were the troops informed of the true nature of their mission. The news came as a stunning surprise to most of them despite the evidence of increased concentrations and other indications of a planned campaign; even in June most had been convinced that their presence in the east was 'one of Hitler's large-scale deception measures designed to hold the Russians in check'.[48]

Despite further evidence, including detailed (and accurate) operational and tactical information supplied by Sorge, despite the evacuation of dependants from the German embassy and the burning of the archives in the basement, and despite reports from the Western Special Military District of preparations to its front, Stalin flatly refused to believe the danger.[49] Instead, he hoped that a summit meeting could be arranged, the promise of which was a false trail cleverly laid by Hitler when the Soviet Foreign Minister, Vyacheslav Molotov, visited Berlin in November 1940. With this in mind, a communiqué was issued by the Tass news agency on 14 June. The total lack of response still failed to stir Stalin, although even Dekanozov was now convinced and presenting him with forecasts of imminent invasion.[50] Finally, on the evening of 18 June a German deserter claimed that he had struck an officer, that his father was a Communist and that he feared for his life. He stated that the invasion would begin at 0400 hours on 22 June. After three days Zhukov received word from the Kiev Military District of the desertion and reported it to Timoshenko and Stalin, who summoned them to the Kremlin. Stalin had also received word through Golikov from sources in Sofia and from Sorge and naturally appeared worried. A directive was issued warning of impending attack but crucially on Stalin's express instruction no other measures were to be taken and 'provocation' was to be avoided at all costs. As Barros and Gregor so succinctly put it, provocation was 'the buckle of Stalin's belt of erroneous assumptions'. When that night another deserter confirmed the report of the first, frantic efforts were begun to convene a summit and resume the negotiations that the Nazis had so studiously drawn out over the preceding weeks.[51] Early the next morning, guns roared all along the frontier and the greatest war in human history began.

The choice of 22 June for 'B-Day' was propitious. As it was a Sunday, many Soviet personnel were on weekend leave. Nevertheless, indiscretion or indiscipline by the troops moving to within metres of border obstacles, bridges and other sensitive points might have alerted the Soviets to impending doom, had Stalin's timidity not ensured that the borders were utterly unprepared. By the end of the first day's fighting the Soviets had lost around 2,000 aircraft, mostly destroyed on the ground (it would take the Red Air Force two years to recover), and German spearheads were up to eighty kilometres beyond their start lines. After a week of colossal losses the pressure began to tell. While leaving the Defence Ministry with Molotov and others, Stalin loudly proclaimed: 'Lenin left us a great inheritance and we, his heirs, have fucked it all up!' Molotov stared at Stalin in amazement but, like the others, wisely said nothing.[52]

If the successful achievement of surprise by the Germans seemed remarkable before, then the overwhelming evidence that was available in plenty of time makes the story utterly astonishing. It would be easy to put it down to the blind stubbornness and self-deception of Stalin and a system in which all power was concentrated in his hands, yet barely six months later the Japanese achieved an equally astonishing coup with no less indication of what was about to happen.

DEVELOPING STRATEGIC DECEPTION: THE LONDON CONTROLLING SECTION

What made British deception during the Second World War unique was its steady development. From an expedient started in the days when things appeared unremittingly bleak, it grew into a flexible and highly effective instrument capable of greatly enhancing operations at all levels. The story of this development was not seamless or without its problems, but the idea that a policy of aggressive strategic deception was both desirable and possible was essentially down to one man, Wavell.[53]

The keys to success were double agents and the false order of battle. While both of these elements were being developed under the wing of GHQ Middle East during the early war years, a double-agent system was being created simultaneously in the UK itself, but crucially this remained a 'private army' for a long time, which nearly destroyed its deceptive value in 1943. It started before hostilities began, with a Welsh-born engineer called Owens travelling on a Canadian passport, who was in contact with both the British Secret Intelligence Service and the Abwehr. During 1939 the Abwehr sent a radio transmitter to SNOW (as he was christened by the British), which he duly handed over. When war broke out, however, the authorities took no risks and SNOW was imprisoned in Wandsworth, from where he began transmitting to Germany under the guidance of section B1a of MI5.* Apart from security matters, MI5 was faced with the problem of what information to send, a problem overshadowed throughout 1940 by the peril in which the country stood. In September a meeting was held between the Directorate of Military Intelligence and MI5, and referred to the Chiefs of Staff Committee. For a while information was passed to emphasize the strength of the defences should an invasion occur. Since this obviously also included naval and air matters and once started would need careful control, in January 1941 the W Board was created, which consisted of the three directors of service intelligence together with a representative of

---

* SIS became part of the military intelligence directorate on the outbreak of war and has been known as MI6 ever since. The Security Service became MI5.

MI5 and Ewen Montagu, then on the Director of Naval Intelligence's staff, to act as secretary.

This happened without specific authorization from anybody. The board reported to no one and was responsible to no one. Yet in order to function and particularly to pass information that might be true, it would require clearance from the Chiefs of Staff which might either be refused or delayed. At its first meeting on 8 January 1941 the W Board decided to institute the Twenty Committee (from the Roman numerals for twenty, a double cross), which would have the executive responsibility to select and approve the material to be sent, guided by the W Board. The chairman they appointed was Sir John Masterman,

> a gaunt humourless man, [who] allowed himself to become obsessed with intelligence and his own importance in it, lord it over many of his wartime colleagues and quite a number of professional regular staff as well, and finally to defy the rules and publish a book about it.[54]

At its first meeting six days later the Twenty Committee discussed the problem as it appeared to MI5. Having established quite an extensive network of double agents, the committee knew that these would have to supply a significant amount of true information to maintain their credibility, but also that there was significant scope for deception. This began with reports on the effect of German bombing and from then onwards, while the Twenty Committee met weekly, the W Board met increasingly rarely: four times in 1943, twice in 1944 and once in 1945.[55]

While this arrangement accounted for the control of the double agents, there remained no direction of deception policy. Nor was deception initially the priority of the Twenty Committee. Masterman later listed seven benefits from the double-cross system, of which deception came seventh.[56] During 1941 the attention remained focused on air matters, although the Home Defence Executive (HDE) and the service departments found it difficult to supply information to be passed on. In a bid to create something, MI5 organized a ludicrous raid on a food store that nearly ended in complete disaster for all concerned and, while throughout 1941 and 1942 MI5 and the Twenty Committee remained geared to defensive misinformation and the tricky business of keeping the double agents in existence, it was 'A' Force that was developing and refining the techniques of operational deception.

By March 1941 'A' Force had already proved its worth tactically and Wavell recommended that in the light of experience gained a controlling authority be set up in London to co-ordinate deception operations in all theatres of war; each command, however, should have its own deception unit. Dudley Clarke was too busy with the traumatic events in Greece and Yugoslavia, and with the arrival of Rommel in Tripolitania, to meet the Chiefs of Staff Committee before October 1941, but he proved persuasive. On 8 October the Joint Planning Staff (JPS) recommended that an organization along the lines of that in Cairo be set up.[57] The ex-War Minister, Colonel Oliver

Stanley MC, was appointed 'Controlling Officer' and Head of the Future Operational Planning Section of the JPS. Stanley believed opportunities for deception must be awaited rather than created, and as a result he came in for criticism from Wavell (now Commander- in-Chief India), who believed that deception should be bold, imaginative and proactive.[58] In any case, Stanley was severely handicapped from the beginning by lack of support from the services and the almost complete absence of experience of strategic deception in Whitehall. The hopelessness of his position was demonstrated by his being permitted to know nothing about MI5's double agents, only that the means existed to plant information on the enemy.

The first attempt at strategic deception was a notional assault on Norway approved in December 1941, called Operation HARDBOILED. It gave the staff some practice and served as cover for the units training for a real operation, the invasion of Madagascar in May 1942. This also gave 'A' Force the opportunity to threaten the Dodecanese islands of Kos and Leros with notional forces based on Cyprus. The landings on Madagascar achieved complete surprise, although the deception authorities could not claim a proven success, only that a plausible alternative was available should the departure of the expedition be blown.[59] But problems continued and half-way through 1942, finally frustrated by an inept attempt to involve SOE*

---

* Special Operations Executive, a secret service formed under the political control of the Minstry of Economic Warfare to promote resistance and subversion within occupied Europe.

agents in deception plans, Stanley resigned his post to return to politics. While Stanley was being replaced by his deputy, Lieutenant- Colonel John Bevan MC, Wavell sent a memo to the Prime Minister recommending 'that policy of bold imaginative deception worked between London, Washington and Commanders in the field by only officers with special qualifications might show good dividend'.[60] Churchill followed this up and the JPS proposed that Bevan's section should be responsible for deception globally and known as 'The Controlling Section', concentrating on broad policy and co-ordinating theatre deception plans. The Chiefs of Staff approved these suggestions in their entirety and the London Controlling Section (LCS) came into being.

Bevan was an old Etonian and head of a respected stockbroking firm who had joined the Hertfordshire Regiment (Territorial Force) in 1911 and had a distinguished record during the First World War. He had been involved in tactical deception during the ill-fated Norwegian campaign of April 1940, was eminently practical and knew – or could get to know – everyone that mattered. He was undoubtedly a first-class choice for the position as head of the LCS, but rather than receiving a qualified staff officer with experience of both operations and intelligence as a deputy, Bevan got a committee that included an Indian civil servant, an actor, a soap factory manager and the novelist Dennis Wheatley. The myth has since grown up that deception was the product of this group of 'gifted amateurs'. Wheatley later wrote that 'by threats and ruses we had kept 400,000 German troops standing idle, in readiness to repel attacks that never matured. Not a bad performance for seven civilians.'[61] However, the credit rightly belongs to soldiers, notably Wavell, Clarke, Bevan and Wild. LCS was not properly integrated into the staff structure and had no control over the Twenty Committee or any of the other agencies upon which it depended for implementation of its schemes. At one point Bevan complained that 'no one tells us anything or gives us any orders', and 1943 would show the limitations of this arrangement.[62]

In fact, only the Controller himself and his deputy, Sir Ronald Wingate, another ex-public schoolboy, were really effective. In the 'old boy' network atmosphere that prevailed, much vitally important business would be conducted after luncheon or dinner in clubs and restaurants.[63] But what they needed at this time was a review of all the means available, including double agents; from this, action could be taken to support the genuine plans emanating from London, in order to co-ordinate them with operational plans elsewhere. The false order of battle should have been started immediately and that work needed reliable, professional soldiers. Eventually, this would come to pass, but in the meantime it was 'A' Force that remained the repository of deceptive wisdom. Bevan remained in constant touch with Clarke and the two worked closely together, so that devices and plans worked out in Cairo were eventually employed on every front. Wavell had Major Peter Fleming (brother of James Bond's creator, Ian Fleming) spend a few months learning the ropes in Cairo before starting a deception unit for GHQ New Delhi. Once

Eisenhower set up his headquarters in Algiers, Clarke sent Major Michael Crichton to run deception there with Colonel E. C. Goldbranson, US Army, and when the deception plan for the Normandy invasion required drafting at Supreme Headquarters Allied Expeditionary Force (SHAEF), it was Noël Wild who took the post. The influence of 'A' Force became global, and a composite weapon was forged that far outdistanced anything even Wavell might have envisaged when he first summoned Clarke to join his staff in 1940.[64]

The seed of strategic deception in the Middle East had been sown when Brigadier John Shearer, Director of Military Intelligence in the Middle East and founder of SIME, returned from Britain in the autumn of 1941 to find that a parachutist had been captured following an air raid on Haifa. The man was German and claimed to be part-Jewish, and said that, having been seconded to guide Italian bombers based on Crete, he had taken the opportunity to try to reach relatives in Jerusalem. The man did not know that a radio he had brought with him had also been discovered and suspicions duly aroused. It transpired that he was in fact the gauleiter of Mannheim. Soon afterwards, an officer experimenting with the transmitter received an answer from Bari. This was reported to Shearer while the gauleiter languished in a prison camp in Palestine, and it was not long before a notional agent was operating as the real agent code-named GAULEITER. Being part-Jewish and an English-speaker, he was given a 'job' as a steward in a senior officers' mess at GHQ Middle East, where he could 'overhear' conversations and he would be freed from having to answer potentially embarrassing questions since nobody had 'talked about it in the Mess'. Since he could not answer specific questions put to him by his controller in Bari, it occurred to Shearer that he could 'overhear' snippets of a cover plan for the forthcoming British offensive Operation CRUSADER, the preparations for which were impossible to conceal.

With the Nazis reaching deep into the USSR and threatening the oil fields around the Caspian Sea, GAULEITER was ideal for reporting radical and alarming changes in British policy. After it had been cleared through the senior commanders of the three services and the Minister of State in the Middle East, the cover plan was put out that a great move northwards was to be made by Ninth Army, commanded by General Sir Henry Maitland Wilson, and that this was being heavily reinforced from the Western Desert while preparations for an offensive there to relieve Tobruk were themselves no more than a cover plan. So successful were the various efforts that Rommel was in Athens when Eighth Army crossed its start line on 18 November.[65]

During the following spring, as blow after blow rained down on the battered Eighth Army during May and June, the CHEESE network kept up reports of preparations in Cyprus for a diversionary descent on Crete. These produced a series of reconnaissances, first by the Regia Aeronautica (Italian Air Force) and later by the Luftwaffe itself, of the airfields not only in Cyprus but also in Lebanon, and of landing craft concentrated in the harbours of Famagusta, Limassol and Larnaca. At the same time a frenzied wireless traffic was maintained between Wilson's headquarters and the 'corps' in Cyprus and its notional brigades, and between the Admiralty in Alexandria with Haifa, Beirut and the Cyprus ports.

In June 1942, when Eighth Army was in great peril, the only restraint on Rommel's advance was the constriction of his supplies from Italy by the submarines and aircraft based on Malta. Malta in turn was under constant attack, and if she was to hold out it was imperative that she also be resupplied. The need to ease the pressure on Malta was acute from April and from that time 'A' Force began building up 'landing craft', 'aeroplanes' and 'tanks' on Cyprus while the CHEESE network began reporting the build-up through a sub-agent in Cyprus and piecing together small items of information indicating offensive preparations against the Greek islands. Although enemy intelligence was interested, according to Clarke the operation was a failure nevertheless. The enemy did not react in the manner wanted; no forces were diverted, nor were they the following month when similar measures were aimed at slowing Rommel in Cyrenaica.[66] None the less, the pieces were slowly coming together when three men arrived by submarine in Palestine, shortly after the Crete invasion scare, and were immediately captured. The new arrivals, all Greek, were led by the man who became known as PESSIMIST. Another Greek double agent, also with two companions, arrived by caique and became QUICKSILVER. Their contacts included a woman called GALA, who was notionally a high-class prostitute in Beirut, and a thug who 'served' on a Greek destroyer. Together with the wholly notional agents HUMBLE and ALERT in Syria and

LEMON in Cyprus, the pieces were in place for the 'Balkan Invasion' that would form the basis of the strategic cover plans for 1943 and 1944, when the Allies took up the strategic offensive.[67]

In 1943 Germany's strategic problem was diffuse: the allies might launch a cross-Channel invasion of Europe or, having cleared the Axis from North Africa, launch a knockout blow against a tottering Italy. Matters had not been improved by Hitler's declaration of war on the USA following the Japanese attack at Pearl Harbor. Barely had the British achieved a breakout at El Alamein than another powerful Allied force had landed at Casablanca, Oran and Algiers in Operation TORCH. To cover this new expedition a notional threat was mounted against the French coast with hardly any resources, but deriving considerable benefit from the ill-fated Dieppe raid of 19 August, which had put all the coast defences on high alert. A notional attack on Norway – SOLO I – was also mounted and a deception practised on the assault troops themselves, called SOLO II. This said that the assault would be made on Dakar. Operation TOWNSMAN provided cover for the all too visible preparations at Gibraltar with a cover story that they were part of Malta relief operations. But this did not account for the landing craft and other amphibious equipment and the LCS was wary of playing up the Dakar story too highly in case of alerts by the Vichy French or the Kriegsmarine, the German Navy. In Barcelona and the Vatican indiscreet inquiries were made regarding Sicily, and the result of all these efforts was the maintenance of security and the achievement of complete surprise.[68] The Allies were still some way short of true strategic deception, but 1943 would provide the opportunities necessary to hone their techniques while demonstrating dangerous weaknesses in London. Operations BARCLAY and COCKADE clearly demonstrate the difference between 'A' Force operations – closely controlled and co-ordinated from the highest level while working with the full co-operation of the Chiefs of Staff – and those of LCS – reliant on the co-operation of local Chiefs of Sections, who were not part of the organization and, perhaps unsurprisingly therefore, did not get the backing they required.

At Casablanca in January 1943 important decisions were taken on future operations that put the emphasis on the occupation of Sicily, while in April planning started for the eventual cross-Channel invasion following the formation and tasking of a staff under Major-General Sir Frederick Morgan as Chief of Staff to the Supreme Allied Commander Designate (COSSAC), whose instructions included preparing 'an elaborate camouflage and deception scheme to pin down the enemy in the west and keep alive his expectations of attack in 1943'.[69] This immediately presented Bevan with a serious problem, since at this stage no genuine cross-Channel attack was planned. CASCADE, the order-of-battle deception in the Mediterranean, had as yet no counterpart in Britain, and without genuine or notional forces to create a threat the Germans were unlikely to take one seriously. Included in Morgan's staff was a section responsible for deception, called Ops 'B', under Lieutenant-Colonel John Jervis-Reid. Together with the LCS, Ops 'B' drew up a plan comprising three distinct elements under the umbrella code-name of COCKADE. The first element was STARKEY, a notional attack by fourteen British and Canadian divisions to establish a bridgehead around Boulogne with a D-Day of 8 September. (This was a serious mistake: a strategic deception plan can never have fixed timings since it will never take place.) This would involve a naval demonstration. Three weeks later an American corps would notionally sail from Britain to capture Brest, under the code-name WADHAM, and finally, since this would be 'called off' for various reasons, five divisions would seize Stavanger in Norway in Operation TINDALL.[70]

A false order of battle was now implemented, somewhat belatedly. Operation DUNDAS was supposed to exaggerate the British and Canadian forces, while operation LARKHILL did the same for the Americans, although a shortage of signalling equipment meant that there was little corroborative evidence to act as backup. Fortunately, however, once established in the minds of Fremde Heere West (FHW, the intelligence branch of OKH responsible for the theatre), these forces were usually accepted as genuine. Based at Zossen, near Berlin, under the command of the patrician Oberst Alexis, Baron von Roenne, FHW finally reported in October 1943 that the Allies had available for a landing no fewer than forty-three divisions, when the real number was just seventeen.[71] But these 'divisions' were established too late to save COCKADE. Without forces, real or notional, to mount the threats, the threats lacked all credibility and foundered; in so doing, they risked exposing the double agents that would be so vital to the real invasion the following year. Over a dozen sources were used by the Twenty Committee to put across COCKADE, but by far the most valuable were TRICYCLE and GARBO. In July TRICYCLE was allowed to go to Lisbon to meet his Abwehr controller, an extremely risky thing to do. Fortunately, he returned with a detailed questionnaire mainly regarding industrial production and orders of battle and he was able to reingratiate himself with Berlin. GARBO, a Spaniard called Juan Pujol who had a large network of notional sub-agents, including a clutch of 'Welsh nationalists', sent a stream of reports which included the date of 8 September for the cross-Channel operation. In due course he was forced to send a message announcing its cancellation, which could easily have destroyed his credibility. It was to prove extremely fortunate for all concerned that this did not happen.[72]

Repeated attempts were made to lure the Luftwaffe into battle over the Pas de Calais, but with no success. Even when a flotilla of thirty vessels assembled near Dungeness and sailed to within ten miles of the French coast, the Germans did not take the bait. A real invasion would be a colossal undertaking, and they were not going to be drawn by anything less. At no stage was OKW deceived as to Allied strategic plans. On 11 July it informed Gerd von Rundstedt, then Oberfehlshaber West (OB, Commander-in- Chief West), 'the *schwerpunkt* of the enemy attack on the mainland of Europe lies in the Mediterranean and in all probability will remain there.'[73] Although Michael Howard suggests that von Rundstedt's staff drew 'highly alarming' conclusions from the whole experience, the report he quotes does not give an impression of fear or panic.[74] Apart from suitable precautions against raids, Calais and Caen were left with virtually no reserves and between 23 June and the end of STARKEY German forces in the west were actually reduced from forty-five to thirty-five divisions. This number began to rise again only after October.

The result was that Kesselring, who was now OB South, had plenty of troops to make a stand south of Rome, thus committing the Allies to a long and difficult campaign in Italy and, although von Rundstedt remained nervous about long-term developments in Britain, his own staff noted that 'the general makeup and number of agents reports gives rise to suspicions that the material was deliberately allowed to slip into their hands'.[75] In other words, the double agents were coming dangerously close to being blown: there were too many of them and some of them were dangerously high-grade. In contrast, rigid selectivity was always exercised in respect of Mediterranean agents to be used for deception, and many were rejected for being too heavily involved with neutrals or undesirable allies. Out of a very large number arrested, only about a dozen were ever used to form the basis of spy rings, and of these all but one was under permanent restraint and at least four were completely notional. The strict enforcement of the false order of battle with bogus divisional signs prominently displayed resulted in casual travellers doing the work and helping rather than hindering security.[76]

By the same token, WADHAM came to absolutely nothing. The US VII Corps, supposedly due to launch the 'attack', found itself unable to co-operate with the Navy until two weeks after the notional D-Day, and Brittany remained steadfastly devoid of German troops. The enemy had been convinced by aerial recce 'and doubtless by other sources' that the invasion of Europe was not yet due.[77] TINDALL was no more successful. 'Judging by the lack of enemy recce which this operation was designed to achieve, it would appear that the operation was a failure.'[78] Having begun July with shipping movements, airfield displays and radio traffic, TINDALL was stood down before it actually had to do anything, ostensibly to release forces for STARKEY. An FHW report did refer to a threat of between four and six divisions on 29 August[79] and the garrison of Norway never fell below twelve divisions (albeit static low-grade formations), but any suggestion of success can be attributed to Hitler's intuitive sensitivity to threats to Scandinavia.

The organization of deception in Britain had proved dangerously weak. Even when deception had been made an intrinsic part of the operational plan, the various bodies involved in it were in no way subservient to its requirements. Up until the end of 1943 the LCS had absolutely no control over the various deceptive components (Bevan later described his position as a 'tinpot pedestal'). The Twenty Committee used the same agents for deception as for counter-espionage and risked their subversion or exposure. This was only resolved later when control was transferred to Ops 'B' at SHAEF, making it the operational deception organization in Britain and leaving LCS to co-ordinate plans worldwide.[80]

In the Mediterranean, by contrast, a long-standing false order of battle together with genuine operations created scope for deceptive ones; a threat to the south of France could be created by the forces preparing in North Africa while Italy could be written off as a dead end, and the threat to the Balkans from Ninth, Tenth and Twelve Armies could be maintained. Ever since TORCH, the Hungarians and Romanians had been convinced of Allied plans to land in the Balkans. This was a view they pushed forcefully to the Germans and which found favour, while the continuing success of Josip Broz Tito's partisans in Yugoslavia had led to the dispatch of a British military mission, focusing attention yet further on the Balkans. These developments permitted the British to disguise Sicily as the real objective by bringing the false order of battle into play with notional assaults on Greece, Sardinia and Corsica. Following discussions between Clarke and Bevan, 'A' Force continued to operate as before, with general guidance from the LCS. There was also the added bonus of the French joining the Allies, under Admiral Jean Darlan, which meant the addition of the Deuxième Bureau and additional double agents under its control, including WHISKERS, LLAMA and, most importantly, GILBERT, who was ultimately to prove as valuable as GARBO.[81]

Operation BARCLAY was created from direction given by the LCS in April 1943, but its detail was planned by 'A' Force. Its principal aim was to secure maximum surprise for Operation HUSKY, the invasion of Sicily due to take place in July. It was also intended to weaken the garrison and retard its reinforcement, particularly by German troops, and to draw off air and naval forces that might interfere with the operation. This was a difficult and subtle task because, as Churchill said of the real objective at the time: 'Anybody but a damn fool would know it was Sicily.' The deception was aimed at containing enemy forces in Sardinia, Corsica and the southern Balkans by simulating preparations to attack these areas from North Africa and the Middle East, and also to encourage the enemy to believe that Allied amphibious operations would only be conducted in periods of no moon.[82] The cover story was that this would be the real 'Second Front', but in order to draw off German troops secondary assaults would be launched against southern and northern France by Alexander, while Lieutenant-General George S. Patton would by-pass Sicily and Italy, and assault Sardinia and Corsica. The notional British Twelfth Army would assault Crete and the Peloponnese, supposedly at the end of May, but the vital element was postponement on grounds that would be believable, since if things were built up and nothing happened, enemy vigilance would relax. In fact there would be two 'postponements', the second of which would take the notional D-Day to the end of July, beyond the genuine one.[83]

As it was impossible to persuade air force commanders to devote many resources to the diverse targets necessary to draw off enemy attention, and with preparations going ahead in Tunis and Bizerta, these were always going to be a difficult objective to achieve, especially when on 11 June the island of Pantelleria was captured with 11,000 Italian prisoners and one British casualty (a soldier bitten by a mule). A major radio deception was created, based largely on Twelfth Army in Cairo, and Victor Jones turned 74th Armoured Brigade into '8th Armoured Division', while Tobruk was filled with dummy landing craft together with a genuine anti-aircraft brigade, fuel and other logistic installations. This provided the background for Major Martin's misfortune, and what Bevan called 'the crucial MINCEMEAT letter' told the Germans what other sources already led them to believe. The Abwehr passed the documents to FHW (who were usually very sceptical about the former's information) and FHW passed it to the Wehrmachtführungsstab (OKW operations section). Throughout the two months following the discovery of Major Martin, OKW continued to give maximum priority to Greece.

The measure of BARCLAY's success is that between 9 March and 10 July the total number of German divisions in the Balkans rose from eight to eighteen, and in Greece from one to eight.[84] The formal inquest declared that it was 'the largest exercise in systematic deception yet attempted in the Mediterranean theatre'. Captured documents and prisoner interrogation on Sicily showed the totality of the surprise achieved, a measure assisted by the genuine movements of the invasion fleets which, supposedly targeted at Greece, converged in the general area of Malta and suddenly turned north in the darkness.[85] Indeed, such was its success that, as the Allied invasion fleet approached in the rough and unfavourable weather, the Italian admiral in charge of the coastal defences was woken with news that an armada was in the narrows: 'Well,' he said, 'at least they aren't coming here,' and went back to sleep.[86]

The success of 'A' Force was due principally to the fact that eighty per cent of the work had already been done in the painstaking assembly and selling to the Germans of the three almost entirely bogus armies of CASCADE. This meant that in Paiforce (Ninth and Tenth Armies in Persia and Iraq) fully ten divisions were concocted, including 31st Indian Armoured Division and three from the Polish Army of the East. CASCADE, in fact, comprised isolated (and often true) items of information provided by the military, naval and air forces and the various Thirty Committees (the Middle Eastern versions of the Twenty Committee, each chaired by an 'A' Force representative and based on Cairo, Beirut and Baghdad) in order to give a misleading overall picture to the enemy. For example, the assembly of landing craft in Egypt with particularly large concentrations at Alexandria and Tobruk fell to Thirty and Thirty-One Committees; airfield preparations in Egypt, Syria and Cyprus to Thirty-One Committee; and the arrival of 56th (London) Division in Iraq from the Western Desert in transit to Syria to Thirty-One and Thirty-Two Committees. Following such items as these would come low-grade observations and gossip, such as the fears of Cairo taxi drivers that more Aussie troops were expected from Palestine, the arrival of Polish officers from Persia and the arrest of an officer for trying to change special Greek money in a bar in Algiers.[87]

To maintain credibility in a long-running deception, it is necessary to possess an escape clause. The constant failure of Ninth, Tenth and Twelfth Armies actually to do anything might have been expected to arouse German suspicions and the failure of the Allies to invade the Balkans could have discredited the double agents. But three things ensured the credibility of the plan. The first was the meticulous way that the false order of battle was maintained, particularly by radio traffic. The second was the fall of Mussolini and the Italian surrender, which made the invasion of Sicily appear to be an expedient taken in changed circumstances. (After the landing on Sicily the Germans discovered the code-name HUSKY but ascribed it to an assault on Greece in which the assault forces were to have taken part.[88]) The third was the shipping calculation. However strong Allied ground forces might grow, the limited availability of shipping and especially of landing craft restricted the options available. The Germans therefore believed the notional armies were being kept from action only by a shortage of shipping.

The great strength of 'A' Force's deception came from its being a continuous process, with major cover plans dovetailing into one another and based firmly on the false order of battle which Clarke had inaugurated when he first joined Wavell. Thus the cover plan to relieve Malta, based on a bogus corps in Cyprus belonging to Ninth Army (itself largely bogus), merged into and continued to assist BARCLAY, and BARCLAY would go on to assist BODYGUARD. When there were no longer any real forces to speak of, it was possible to abandon truths for outright lies and still to continue the illusion.[89]

Chapter Seven

NAVAL DECEPTION

'. . . he communed with his counsellors, and all were of one mind to follow up the Genoese, so they hoisted sail and pursued after them. But you must know that they were deceived.'

Martino de Canale

The Age of Sail

Vegetius, writing in the fourth century AD, describes how Roman skiffs used for reconnaissance had their sails and rigging dyed Venetian blue

which resembles the ocean waves; the wax used to pay ships' sides is also dyed. The sailors and marines put on Venetian blue uniforms also, so as to lie hidden with greater ease when scouting by day as by night.[1]

Warfare at sea has obviously been subject to bluff and deception for as long as warfare on land. In 1264, during the long wars with Venice, the Genoese decided to intercept the 'caravan of the Levant', an annual convoy that the Venetians sailed to Egypt and Asia. The caravan was an event of great moment. Its dates of departure and return were fixed by strict laws, as were the numbers of men on each vessel and the conduct of the convoy itself. The commanders and captains were chosen by the Great Council and in times of war the Senate pronounced the *chiusura del Mare* ('closing of the sea'), a decree that forbade any vessel from leaving the convoy, while arrangements would be made to escort it with war galleys. The Genoese well understood the importance of this convoy to Venice and decided to send Simone Grillo with twenty galleys, two large vessels and a contingent of 3,500 men to intercept it. In reply, the Venetians assembled a force of no fewer than forty-seven galleys under 'a brave man and wise, and sprung of high lineage', Andrea Barozzi. This 'noble captain' set out for Sicily expecting to intercept the Genoese before they in turn could attack the caravan.

Alas for Barozzi, on this occasion his wisdom failed him. The Genoese were indeed there, but all he found was 'a boat in which there were men who told him on inquiry that the Genoese galleys had passed four days previously, bound for Syria'. After a hastily assembled council of war, Barozzi set off in a fruitless pursuit and as soon as the news reached Venice orders were given for the immediate departure of the caravan, which had been delayed owing to the supposed presence of the enemy in the Adriatic. Grillo now emerged and put his fleet into position at Durazzo to await the arrival of the caravan, the movements of which he was kept fully informed of by an underwriter of the Great Council (who, the chronicles note with barbed acidity, came from Treviso). When in due course the caravan was intercepted, its commander, Michele Duaro, tried bravado, throwing some chicken coops in front of the Genoese line and bidding them fight the chickens. However, this served no purpose and with no escort of warships the caravan was soon destroyed, as grievous a blow to Venetian prestige as to her material well-being.[2]

Not only does this episode illustrate an early example of deception in naval warfare, but it also shows the importance of commerce to naval strategy. While the principles of warfare and of deception apply equally on land and at sea, there are obviously fundamental differences. While land warfare is fought with units containing thousands of men and hundreds of pieces of equipment, naval warfare is conducted with dozens of units or (more usually) fewer, each of relatively great value. More importantly, it is fought over a vast area, with no natural cover. The size of the ships also makes it hard to conceal or disguise them and their shapes make identification of their nationality and class quite simple, so that deception is difficult – but not impossible. Since it was common in the days of sail to capture enemy shipping rather than to destroy it, it was equally common for foreign-built ships to serve with the navies that had captured them, and therefore not unusual to see them bearing different colours from their country of origin.* Over the years many other measures have been adopted to suggest that a ship is not what it appears, giving plenty of scope for tactical deception.

Thomas Cochrane, tenth Earl of Dundonald, was a daring and inspirational leader who was always prepared to use guile combined with forethought and audacity to overcome large odds, in other words a master of deception. He was convinced (and proved) that a single ship correctly handled, preying on coastal shipping and coast defences, could cause the enemy loss and distress out of all proportion to the effort expended. He took great pains over the training and welfare of his men and this paid dividends in their performance. His first command was the 168-ton brig HMS *Speedy,* which he operated off the Spanish coast in 1800. Knowing the Spaniards would soon come to recognize his vessel for an enemy, he repainted it to resemble the neutral Dutch ship *Clomer,* which had been trading in the area for some time. He also recruited a Danish speaker whom he provided with a Danish uniform. Towards the end of December he gave chase to what appeared to be a heavily laden, unarmed merchantman, only to discover as he drew near that he too had been duped. It was a Spanish frigate with some 200 men and heavy guns, which now put down a boat. He ordered below everyone who looked British, and set his 'Dane' to tell the Spaniards they were neutrals. When this failed to

---

* The French frigate *Résistance* for example, which was captured by the Royal Navy following the landing at Fishguard, was renamed HMS *Fisgard* in the service of her new masters. Even during the Second World War both the Royal Navy and Kriegsmarine put captured submarines into service.

put them off, one of his men hoisted a yellow flag (quarantine) to the foretop and the 'Dane' said they were just out of Algiers. The Spanish knew that Algiers was suffering from an outbreak of bubonic plague and quickly returned whence they had come.

Three months later Cochrane was chased by an enemy frigate, which gained on him throughout the day and was guided at night by the faint glimmer of light from the little brig. But as they drew near towards daybreak, the enemy frigate found it had been chasing a tub with a lantern in it and the brig was nowhere to be seen. Cochrane later used the same ruse again. Commanding the frigate HMS *Pallas* in March 1805, he was chased by three French 74-gun ships of the line off the Azores. After conducting a brilliant manœuvre to run back on them, he was pursued for the rest of the day and all night, but when they closed in for the kill all they found was a ballasted cask with a lantern made fast to it.[3]

Captain Raphael Semmes and the Confederate cruiser CSS *Alabama* forged a formidable reputation as a commerce raider. The *Alabama* sank no fewer than eighty-three US merchantmen as well as the heavier gunboat USS *Hatteras* (which she lured to her doom by pretending to be a merchant blockade runner[4]), and was probably the most famous ship in the world at the time. The USS *Kearsarge* had been pursuing the *Alabama* for a year in European waters when, as she lay at anchor in the Scheldt estuary near Vlissingen on Sunday 12 June 1864, her captain, John A. Winslow, received word from the US minister in Paris that his elusive quarry had steamed into Cherbourg the day before. Winslow wasted no time and two days later found his prey still in Cherbourg roads, where he stopped engines and lay to. Unable to engage within the confines of a neutral port, Winslow retired beyond the three-mile limit required by international law, intending to intercept *Alabama* when she emerged He took precautions against a surprise night attack but was most worried that *Alabama* might try to slip away. The following day, however, he received a note from Semmes via the American vice-consul that indicated his intention to fight at the earliest opportunity and begging Winslow not to depart.

The two ships were evenly matched. Both were three-masted and steam-propelled, and if the *Kearsarge* mounted a combined broadside firing 365 pounds to the *Alabama's* total broadside of 264 pounds, the latter's Blakely guns outranged and were more accurate than the Dahlgrens of the *Kearsarge*. However, the speed and manœuvrability of the *Alabama* were declining and Semmes had intended to put her into dry dock for two months and thoroughly clean the keel and overhaul the boilers. Nevertheless, he wrote in his journal that 'the combat will no doubt be contested and obstinate, but the two ships are so evenly matched that I do not feel at liberty to decline it.' He had confidence in the 'precious set of rascals' that was his crew. Besides, his luck had never yet failed him and he busied the crew preparing the ship, waiting for Sunday, which he deemed his lucky day.

Sunday dawned bright, clear and cool and after a leisurely breakfast the *Alabama* was cheered out to sea by crowds along the mole and in the upper windows of the buildings, where a fine view could be had of the forthcoming action. Excursion trains had brought sight-seers from Paris, and Cherbourg was packed with excited crowds shouting 'Vivent les Confédérés!' In a new dress uniform Semmes delivered a stirring oratory to his men before taking station on the horseblock just before the mizzen mast. Then at 1057 hours, with watch in hand, at a range of about a mile, he asked his executive officer if he was ready: 'Then you may fire at once, sir.'

No hits were scored as the range closed to half a mile, when Winslow returned the fire and the two ships began to circle to starboard, firing furiously at each other. A Blakely round scored a direct hit on the sternpost of the *Kearsarge* but fortunately for Winslow it was a dud. A three-knot current bore the ships westward and as it did so so their circles became tighter until the range dropped to about a quarter of a mile by the seventh and final revolution. Once they were on target, the US guns inflicted tremendous damage. At the same time, Semmes watched in horror as everything his own guns fired at the *Kearsarge* bounced harmlessly off the sides, including solid shot. Realizing the desperate state of his old vessel, Semmes ordered full sail for the coast but *Kearsarge* was not to be denied. When Semmes saw the wreckage to which the lower decks had been reduced, he ordered the colours to be struck saying: 'It will never do in this nineteenth century for us to go down, and the decks covered with our gallant wounded.' Captain and crew abandoned the rapidly sinking ship, which went down at 1224 hours, just ninety minutes after she had opened the action.

Only after the battle did Semmes discover that the *Kearsarge* had 120 fathoms of sheet chain suspended from scuppers to waterline, bolted down and concealed behind an inch of planking: he had been fighting an ironclad! Semmes protested this was unfair. 'It was the same thing', he said, 'as if two men were to go out and fight a duel, and one of them, unknown to the other, were to put on a suit of mail under his outer garment.' Perhaps, but Commodore David Farragut had employed the same stratagem two years previously, when he ran past the forts into New Orleans.[5]

STEAM AND STEEL

The development of the ironclad increased the size and cost of ships. At the same time, improved armaments increased the range at which actions were fought and reduced the scope for capture, making sinking a more likely outcome of an action and thus making it increasingly difficult and expensive to replace losses. But losses must be accepted if control of the seas is to be gained and maintained, as it must be if commerce is to flow unhindered. However, the official history of the First World War describes how

> by a strange misreading of history, an idea had grown up that [a fleet's] primary function is to seek out and destroy the enemy's main fleet. This view, being literary rather than historical, was nowhere adopted with more unction than in Germany, where there was no naval tradition to test its accuracy.[6]

On the one occasion the German Battle Fleet did enter the North Sea to fulfil its aim, it achieved a marginal tactical victory over the British (in simple terms of losses) at the Battle of Jutland, but there can be no doubt as to the strategic result of the battle.[7] The British did not deceive the Germans but simply faced them down, and the German Battle Fleet spent the remainder of the war sitting idly in port while the British naval blockade helped squeeze Germany to ultimate defeat. However, British nervousness of the German Battle Fleet forced her to denude some other vital positions of destroyers, such as Dover. Thus the Dover patrol had to rely on bluff to prevent German naval forces operating from the Belgian ports from interfering with the vital cross- Channel traffic.[8]

Meanwhile, Britain herself came perilously close to being squeezed to defeat by Germany's commerce raiders and U-boats during both world wars. An early effort to counter this threat was camouflage paint schemes. Transport and cargo ships were painted neutral blue, grey or sea-green in the hope of avoiding detection for as long as possible. Warships, on the other hand, are not looking to avoid contact but instead require every fighting advantage they can muster, particularly in the early stages of an action. One result was a proposal by an eminent Scottish zoologist, John Graham Kerr, whose study of marine vertebrates suggested that odd patterns of white and grey might help make ships harder to identify. Although the Admiralty circulated his suggestions as early as October 1914, it left responsibility to individual captains and was later shelved. It took further prompting from another painter, P. Tudor Hart, and an RNVR lieutenant, Norman Wilkinson (a marine painter and poster designer who had served in the Dardanelles campaign) who wrote to the Admiralty on 27 April 1917, to create what was known as 'dazzle' camouflage. In poor visibility, at long range or at high speed, this served to hinder an observer's ability to identify a vessel accurately, perhaps long enough to give it a precious advantage. It also made judging the vessel's speed more difficult – very important when trying to fire at long range. Gunnery officers and submarine captains had to 'track' moving ships on calibrated range-finders and periscopes, but the pattern distorted the image and made it harder to secure a hit. Refinements of the same technique included false bow waves to give the impression of greater speed, false waterlines which were designed to inhibit accurate estimation of range, and painting the upper works a lighter colour to blend them with the sky. The effectiveness of the technique was questionable but it raised morale and was therefore retained, mainly for merchant shipping.[9] Nevertheless, during the Second World War the Admiralty Research and Development Section employed the naturalist and artist Peter Scott to develop further patterns.

The vulnerability of shipping to aircraft, demonstrated among other instances by the destruction of HMS *Repulse* and *Prince of Wales* by the Japanese on 11 December 1941, made it essential to camouflage ships from the air. On the open ocean, ships could not avoid being spotted by aircraft in the vicinity. For example, a US aircraft north of Guadalcanal flying at 18,000 feet spotted five destroyers belonging to Rear-Admiral Tanaka's 'Tokyo Express' at a distance of eight to ten miles, and sighting fast-moving warships at greater ranges was not unheard of in good conditions.[10] Attempts were made to design patterns that gave some protection from aerial attack, but these were seldom effective, at least while a ship was at sea. Eventually, technical developments such as radar and acoustic torpedoes made dazzle patterns largely redundant, but they continued in use throughout the Second World War.[11]

If a ship was inshore, by its very nature it might be found if aircraft looked in the bays, rivers and ports. Paint might go some way to protect it in such circumstances, blending it with its surroundings just long enough to put a bomb aimer off, but a photo interpreter could probably identify its class precisely and thus reveal its speed, firepower and cargo capacity. Nets and, where appropriate, cut vegetation might help to make the tell-tale shape of a ship blend in with the shoreline and barges and floating material could be used to break up the characteristic shape of bow and stern.[12]

Another early measure adopted to counter U-boats was the arming of merchant ships in 1915, which was followed by the creation of Q-ships. These were merchant ships armed with concealed guns and torpedoes manned by naval crews, designed to lure the U-boats – which preferred to destroy merchant vessels by gunfire – to a position where they themselves could be destroyed.* The

---

* Their ruthlessness has also been credited with helping to provoke unrestricted submarine warfare by forcing the U-boats to attack submerged, and thus helping bring the USA into the war. The most notorious example was the murder of the crew of *U-27* by the crew of the Q-ship *Barralong* (see A. Coles *Slaughter at Sea: The Truth Behind a Naval War Crime,* London, Robert Hale, 1986).

Q-ships were eventually credited with eleven U-boat kills out of a total for the First World War of 192.[13] During both wars the Germans operated similar ships as merchant raiders. Perhaps the most famous example was the *Atlantis,* commanded by Kapitän zur See Bernard Rogge during the Second World War, one of nine such ships which sank 850,000 tons of Allied shipping and kept the Allies busy for three and a half years. The *Atlantis* logged over 100,000 miles in 622 days at sea and accounted for twenty-two Allied freighters, making her the most successful surface raider of the war. In the course of her wanderings she pretended variously to be the *Krim* (Russian), the *Kasii Maru* (Japanese), the *Abbekerk* (Dutch) and the *Antenor* (British).

Carrying huge stocks of fuel and food, *Atlantis* mounted behind collapsable bulkheads an armament of one 75mm and six 150mm guns and six light anti-aircraft guns, plus four torpedo tubes, mines and a Heinkel He-114 seaplane for reconnaissance. She had a dummy stack and cargo booms and carried a variety of fake foreign uniforms and clothing, male and female, which the crew could use as appropriate. In addition, there was a large supply of paint to change her name and the colour of the superstructure. It is perfectly legal for a ship to operate in this fashion, providing it displays the correct national flag before opening fire, and Rogge adhered strictly to this law, as well as endeavouring whenever possible to pick up survivors, who were treated graciously.

Rogge trained his gunners to shoot out a victim's radio equipment first, which would allow the remainder of his operation to take place in slow time. None the less, a stream of QQQ messages ('I am being attacked by a disguised merchant ship') eventually helped the Admiralty to track him down. The final clue to *Atlantis*'s whereabouts in November 1941 was provided by ULTRA intercepts ordering her to resupply submarines south of the Equator. On 22 November a seaplane from HMS *Devonshire* (sent to nearby Freetown to look for her) sighted a suspicious merchant ship and opened fire while *Atlantis* was in the process of replenishing *U-126*. Rogge tried one last desperate trick. He signalled urgently (and indignantly) that he was the *Polyphemus,* a Dutch ship, then gave the signal RRR: an Allied cipher that an enemy warship was close by. Unbeknown to Rogge, this cipher had recently been changed to four Rs. A new precautionary system introduced by the Admiralty to plot the whereabouts of every single known ocean-going merchantman confirmed *Devonshire*'s suspicions and when word came from Freetown that this ship could not possibly be *Polyphemus,* Rogge and his crew were forced to take to the boats.[14] Afraid of lurking U-boats, *Devonshire* made off, and after a series of extraordinary adventures Rogge and the survivors were eventually picked up by U- boats and returned to Germany.

THE BATTLE OF THE RIVER PLATE

The introduction of air power and submarines radically transformed naval warfare. The submarine was eventually defeated by the convoy system among other things, but the threat posed by the pocket battleships of the *Kriegsmarine* was something very different. The escorts normally assigned to convoys might be able to deal with submarines, but destroyers and corvettes would be defenceless against the 11-inch (280mm) guns mounted by *Admiral Graf Spee* and *Deutschland,* both of which had put to sea before the outbreak of hostilities in August 1939.

Graf Spee sank just nine ships in her short career, but the success of commerce raiders is not measured merely in terms of sinkings. They completely disrupted commerce simply by being at sea and the Royal Navy was forced to deploy no fewer than nine hunting groups to look for her, most of them withdrawn from other theatres. They affected the operations of soldiers and airmen serving as far apart as Egypt and Singapore, and the Allies were reaping the benefits of her destruction as late as 1944.[15] After operations in the Indian and South Atlantic Oceans *Graf Spee* developed engine problems and Kapitän zur See Hans Langsdorff, anxious to increase his score before heading for Germany for repairs, made for the River Plate, where he expected to intercept a convoy. Commodore Henry Harwood, commanding Force G (the light cruisers HMS *Ajax* and HMNZS *Achilles* and the cruiser HMS *Exeter*), anticipated the move. After an eighty-minute battle fought on 13 December, *Exeter* was forced to withdraw and *Ajax* and *Achilles* forced to disengage.

However, *Graf Spee* was also badly damaged and fled to the Plate estuary, where she sought refuge in neutral Montevideo. Both sides now faced dilemmas. An estimated four days were required for the repairs Langsdorff needed, but international law would only grant him seventy-two hours unless the German embassy could persuade the Uruguayan government otherwise. Alternatively, he could make a break for neutral but friendlier Buenos Aires farther up the estuary or out to the open sea. Meanwhile, the British wished to keep *Graf Spee* in harbour for at least four days while reinforcements rushed to the scene. If *Graf Spee* made a break for the ocean, it could easily sink *Ajax* and *Achilles* in the process. In this instance a deception was needed to hide British weakness.

While *Achilles,* the only ship visible from shore, sent a string of messages as if other ships were waiting just over the horizon, British diplomats loudly demanded that the *Graf Spee* should leave within twenty-four hours, to suggest that the British were keen to finish her off.[16] For a few days at least it would be possible to prevent her doing just that by the timing of departing British merchant ships, which could each claim twenty-four hours' grace under the Hague Convention. Meanwhile, rumours were planted by the diplomats in waterfront bars and casual conversation with other diplomats to the effect that the battlecruiser HMS *Renown* and the aircraft carrier HMS *Ark Royal* had left Cape Town on 12 December and would shortly be on station. The appearance of another light cruiser from the Falklands, HMS *Cumberland,* added some credibility to the reports.

Langsdorff was suddenly given brief hope late on 16 December, when he received reports that *Renown* and *Ark Royal* were in fact in Rio de Janeiro, only to be disillusioned when the harbourmaster informed him that a British merchantman had departed at 1815 hours that evening, meaning he could not leave himself until 1815 hours on the 17th. Since his seventy-two hours would be up shortly afterwards and the British squadron offshore would in all probability be able to shadow him and direct the more powerful units to complete his destruction, the following day Langsdorff scuttled his ship and shot himself.[17]

THE INVASION OF NORWAY

On 2 April 1940 Hitler decided to invade Scandinavia, setting the invasion date as the 9th. At midday on 8 April 1940, while on patrol in the Skagerrak, the Polish submarine *Orzel* surfaced to challenge a large steamer that was heading north. When it discovered this was the German transport *Rio de Janeiro*, *Orzel* gave the crew and passengers time to take to the boats before sinking her by torpedo. *Orzel* slipped beneath the surface again as a Norwegian destroyer and fishing boats arrived to pick up the men in the extremely crowded lifeboats. The latter were alarmed to find that many of the men picked up were German soldiers in uniform, who readily announced that they were on their way to 'protect' Bergen from the British. This 'Trojan Horse' transport was one of seven assigned to eight groups organized for the treacherous invasion of Norway, whose neutrality had been guaranteed by Germany. However, this timely discovery (made public by Reuter's that same evening) failed to rouse the Norwegian government, who refused to believe it and failed even to alert the coast defences. Mobilization had not yet been considered and neither did the British Admiralty react effectively, despite its longstanding plans to occupy Narvik so as to deny it – and the iron ore route from Sweden – to the Nazis.[18]

In fact, the British plan had been an important factor in determining the Führer's decision, and it also provided a means of deception to the Kriegsmarine, who were extremely nervous that the Royal Navy might inflict colossal damage on their meagre surface fleet. Winston Churchill, First Lord of the Admiralty at the time, had finally persuaded the Allied Supreme War Council to mine the Norwegian Leads (Operation WILFRED) on 8 April and, anticipating a violent German reaction, a small Anglo-French force was to be sent to Narvik with other contingents to Trondheim, Bergen and Stavanger, in what was known as 'Plan R-4'.[19] British forces were therefore embarking at Glasgow and Rosyth simultaneously with Germans in Hamburg and Kiel. To prevent British interference with their own invasion, the Germans used the British plan as the basis for a feint. The B- Dienst (the German naval code-breaking and monitoring service, which had given the Germans warning of British plans) alerted them to the sailing of the British invasion convoys and a strong force of battle-cruisers and destroyers would sail immediately to threaten it. As expected, the heavy units of the British Home Fleet withdrew northwards to protect their own convoy and left the Germans a free hand in southern Norway.[20]

The German invasion groups went further with their tactical deception, producing detailed orders for

'Deception and Camouflage in the Invasion of Norway'. These instructed naval units to disguise themselves as British, with arrangements 'to be made to enable British war flags to be illuminated'. If warning shots were fired at them, they were to signal: 'Stop firing. British ship. Good friend.'[21] The German force entered the Kors Fjord approach to Bergen at 0200 hours on the morning of 9 April, aided by fog. When illuminated by searchlights and challenged by a patrol vessel, a signal from the *Köln* identified herself as HMS *Cairo*. This seemed to satisfy the Norwegians, and the ships swept on unhindered. At 0430 hours a Norwegian destroyer received a signal in English that 'I am proceeding to Bergen for a short visit,' which proved sufficient to get by the challenge.[22] At the same time the *Hansestadt Danzig* moored at a pier in Copenhagen and discharged a battalion to seize the Danish capital with hardly a shot fired, and the Nazi party organ the *Völkischer Beobachter* printed a banner headline that screamed 'GERMANY SAVES SCANDINAVIA!'[23]

THE CHANNEL DASH

The astonishing success of Operation CERBERUS (the escape of the German battle-cruisers *Scharnhorst* and *Gneisenau* from the French port of Brest), which humiliated the Royal Navy in February 1942, owed much to German deception efforts, assisted by good fortune and British lethargy. The British had no intelligence contingency plan after the fall of France in 1940, and relied partly on Polish sources. One large spy ring under Roman

Garby-Czeriawski had plenty of naval intelligence, but it also developed contacts with the Abwehr and this proved fatal to it in early 1942. The Germans were then able to use the ring to deceive the British over the seaworthiness of *Scharnhorst* and *Gneisenau* shortly before the Channel Dash was made.[24] German naval losses in the Norway campaign had been disastrous and went a long way to scuppering realistic hopes of invading Great Britain. The loss the following year of her most powerful battleship, the *Bismarck,* caught and sunk while running for the sanctuary of Brest, left the remainder of the force with which she had been due to operate effectively trapped there. Hitler decided that something decisive was about to happen in Norway and that the battle-cruisers, together with the heavy cruiser *Prinz Eugen,* should be redeployed there to operate with *Bismarck's* new sister ship *Tirpitz.* He insisted, however, that this must be achieved via the shortest route, straight up the English Channel.

Hitler's naval planners were horrified when handed this order on 12 January 1942. It seemed a suicidal mission since the RAF, which had been using the three ships for bombing practice, would be in a position to hammer them the entire length of the route. Vizeadmiral Otto Ciliax decided to base his plan on the premise that while the British expected him to make a run up the Channel, he would be able to deceive them as to the precise time and route. Security was essential: no radio traffic was sent and the crews were kept strictly in the dark. Since the Dover straits were the most vulnerable stretch, it would be reasonable to time the run to pass through them in darkness but this would mean a daylight departure. Alternatively, a night-time departure could use the cover of a routine training exercise to conceal his true intentions, at least until the following day.

Ciliax accordingly drew up a timetable for heavy air escort and for destroyer and fast patrol boat support. The minesweeping necessary to clear the route had to be undertaken very carefully so as not to alert suspicion, with the last gaps only covered at the final moment. At the same time French dock workers were allowed to see drums of lubricating oil prominently labelled *Lubrifiants Coloniaux* (high viscosity for hot climates) and sun helmets and tropical clothing were ordered from French suppliers. The ship's departure on the night of 11 February could not possibly be concealed, but the story of fleet exercises and gunnery trials in the Bay of Biscay was reinforced by ensuring that the French harbour authorities sent out their target towing vessels to the exercise area, and tugs and net-laying vessels were ready for the ships' 'return' the following day. A hunting party was arranged at Rambouillet on the 12th and many senior officers accepted invitations.

After being delayed by an air raid and when *Scharnhorst* hit a torpedo net, the three heavy ships with an escort of six destroyers turned first north and then east. By daybreak they had been joined by fifteen large torpedo boats and thirty fast patrol boats, while constant relays of fighters provided air cover. Ciliax was right to assume that he could not deceive the British as to his aim, and the Admiralty produced an estimate at the beginning of the month predicting a break-out through the Channel.

But the Home Fleet remained at Scapa Flow while the First Sea Lord, Admiral Sir Dudley Pound, remained sceptical. On 8 February Air Chief Marshall Sir Philip Joubert, Commander-in- Chief RAF Coastal Command, forecast a break-out within the next few days and ordered mines to be laid in the most probable gaps cleared by the Germans, while Commander (later Vice- Admiral Sir) Norman Denning at the Operational Intelligence Centre in the Admiralty persuaded the Flag Officer Submarines, Admiral Max Horton, to station HM Submarine *Sealion* off Brest.

Unfortunately for the prescient Denning, *Sealion* had withdrawn at the crucial moment to deeper water, the first of a series of strokes of luck to assist Ciliax's dash. The RAF patrol detailed to watch the port had been forced to return to base with a defective radar and by the time it was back on station its neighbour had also developed a defective radar and the weather had deteriorated to provide further cover. Moreover, even when fully operational the British radars were very susceptible to jamming. Consequently, the Germans were only spotted by the RAF at 1035 hours on the 12th by a Spitfire off Le Touquet as they were approaching the Straits of Dover, charging along at full speed. Attacks were belatedly launched but all failed, and the only damage the Germans incurred was caused by Joubert's mines.[25]

It was a fantastic result for the Germans, or at least so it seemed. When Pound telephoned Churchill to say so, the prime minister barked 'Why?' and threw the phone down. In reality, the only benefit derived was propaganda for the surface fleet. Ten days after reaching safety *Prinz Eugen* was torpedoed off Trondheim and she took no further part in the war. At the end of the month *Gneisenau* was pummelled by the RAF for three nights and similarly put out of action (she ended the war as a blockship in the Baltic). *Scharnhorst* was eventually lost with most of her crew in the Battle of North Cape on 26 December 1943. Großeadmiral Erich Raeder said of the Channel Dash, 'it was a tactical success but a strategic defeat,' and in that sense it might be described as the nearest approximation to the Battle of Jutland in the Second World War.[26]

St Nazaire

When *Bismarck* made her doomed break into the Atlantic, it was intended that she would take refuge in France. The Admiralty was determined that her sister ship *Tirpitz* should be prevented from doing the same. The only port with facilities capable of supporting *Tirpitz* was St Nazaire, which possessed an enormous lock that could be converted into a dry dock by pumping out the water. The operation was entrusted to Vice-Admiral Lord Louis Mountbatten's Combined Operations Command, which decided to put the lock out of action by ramming it with an old ex-American destroyer, HMS *Campbelltown,* disguised as a German *Möwe*-class torpedo boat and loaded with explosives. The operation would involve an approach of over 400 miles by sea and sailing five miles up the Loire estuary.

The attackers set out from Falmouth on 26 March 1942 and comprised 353 naval and 268 commando

(Army raiders) personnel on *Campbelltown,* sixteen launches and a motor torpedo boat, escorted by another four destroyers. They narrowly averted interception when they were reported by *U-593* on the afternoon of the 27th, and reached the mouth of the estuary, where two escorting destroyers were standing guard. With German ensigns flying, *Campbelltown* began to move carefully between the mudflats when she was challenged from the shore. A British seaman disguised as a German petty officer managed to signal back sufficient genuine German code to enable them to get past the heaviest German batteries and to reach within two miles of the target before the cover was blown. As the coastal defences opened fire in earnest, the German colours were lowered and the White Ensign run up, but the defenders were unable to prevent the ramming of the lock gates seven minutes later at 0134 hours (just four minutes behind schedule). The commandos rushed ashore to create maximum havoc and the crew were taken off. The fighting throughout the town was intense and one by one the launches supporting the assault were sunk. As they tried to escape, the survivors ran into German torpedo boats returning to the estuary and, although these were driven off by the supporting destroyers, only four of the launches ultimately survived. British losses were 144 killed and around 300 taken prisoner. But ten hours after she had been lodged against the lock gates *Campbelltown's* three tons of explosive blew them (and a large number of curious German officers) to oblivion. On the 29th destruction was completed by delayed-action torpedoes fired from the lost *MTB-74,* which exploded thirty-six hours hard-pressed USSR. Not surprisingly, attempts were made to protect these convoys through deception, and these show the versatility of the double agents. Much of the deception practised by the Royal Navy during the Second World War was of a strategic nature, aimed at painting a false picture of Britain's naval strength and largely prompted by the Director of Naval Intelligence, Rear-Admiral John Godfrey. Initially, some merchant ships in Scapa Flow were disguised as R-class battleships and the aircraft carrier HMS *Hermes*. Later, however, the double agents were the main deceivers and the Japanese the main target.[29] A shipbuilding programme was announced, as a result of which the Commander-in-Chief East Indies acquired two notional aircraft carriers in 1943, when in fact he had none. In 1945, when U-boats were being fitted with snorkels that enabled them to recharge their batteries underwater, the patrolling efforts of the Royal Navy and Coastal Command were proving less successful at flushing them out. The only effective means of countering them was the laying of deep minefields over which surface craft might safely pass. Unfortunately, shortages of both mines and minelayers made this impractical and so the Admiralty enlisted the aid of the Twenty Committee. Double agent TATE was able to use a previous 'contact', notionally serving in HMS *Plover*, to pass a stream of information about new minefields, and this was made more credible by adding details of known U-boat kills before the details of these losses were known to the Germans. When a U-boat struck a mine and was forced to scuttle itself (by great good fortune, in the general area of one of TATE's 'minefields'),

Vice-Admiral William 'Bull' Halsey's Task Force 16 was ordered to leave the Solomon Islands in May to return to Pearl Harbor, with secret instructions to reveal itself to Japanese patrol planes operating with the invasion force heading for Nauru and Ocean Islands to the east. Halsey's two carriers were duly reported on 15 May, whereupon the Americans withdrew, leaving the South Pacific temporarily open. The sighting of the two carriers was sufficient for the Japanese to cancel the proposed invasion of Nauru and Ocean and presented the opportunity for a radio-based deception to begin later that month. Two ships, a cruiser and a seaplane tender belonging to Vice-Admiral Herbert Leary, Commander of Naval Forces Southwest Pacific, based at Melbourne, broadcast traffic modelled on that during the recent Battle of the Coral Sea while cruising off the Solomons and were answered by Leary's shore facilities. The Japanese could only conduct traffic analysis and it appeared to them that Halsey's forces remained where it had been sighted. On 4 June the naval staff in Tokyo informed the Midway force that the Americans remained east of the Solomons, apparently unaware of Japanese plans. This was despite conflicting radio traffic evidence suggesting a force had departed Pearl Harbor for Midway on 30 May.[33] The failure of Japanese intelligence and the use of deception enabled Nimitz to ambush the Japanese strike force at Midway. He went on to sink four of their carriers and wrest the initiative from them once and for all.

Only with reluctance did the US Navy overcome its distaste for deception in the Pacific theatre, where it

[H]ewitt, Eisenhower and indeed Roosevelt himself saw the results the British were achieving and they were in close, friendly relations with the British, so there was some fairly high-powered support, which won the day in the end.[35]

The Beach Jumpers (the name was adopted as a cover) operated from 63-foot Air Sea Rescue boats that were specially equipped with a variety of deception devices, including balloon-mounted conical radar reflectors, loudspeakers linked to wire recording sound systems known as 'heaters', and bombardment rockets. Their training – for which there was no established precedent – included seamanship, gunnery and the use of pyro-technics (including smoke pots and demolitions), as well as courses on electronic counter-measures.[36] Their first operational deployment came during Operation Husky and was deemed a success. A German news broadcast on 13 July reported the repulse of a landing attempt between Sciacca and Mazzara del Valo, and subsequent prisoner interrogation and captured docu-ments suggested that a German reserve division had been held in place there while uncertainty existed as to where to deploy it.[37] With this initial success Fairbanks, who thoroughly understood the principles of decep-tion, was appointed Hewitt's Special Operations Officer, and immediately set about establishing liaison with Eisenhower's Allied Force Headquarters (AFHQ) and, most significantly, a very close working relationship with 'A' Force.

The Beach Jumpers were less successful in their operations to cover AVALANCHE, the Salerno landings, and it was found that their equipment and set-up were inappropriate to the island- hopping campaign in the Pacific. Their most significant operation proved to be BIGOT, which provided cover for the ANVIL-DRAGOON operation in the south of France. There they simulated amphibious threats to areas away from the landings and were engaged by two German corvettes. Destroyer and gunboat support resulted in the sinking of these but more satisfying was the evidence of the success of their deception. Throughout the D-Day of 15 August 1944 Berlin radio reported landings west of Toulon and to the east of Cannes, and that 'thousands' of dummy parachutists had been dropped, adding that 'this deception could only have been conceived in the sinister Anglo-Saxon mind'.[38]

Since 1945 there have been no major naval combats, only skirmishes using 'pieces' of the new technology that has seen a transformation in naval warfare through the introduction of sophisticated sensors and guided weapons. Ships no longer rely on armour or speed for protection, but they do now possess the means to defend themselves by shooting down attacking missiles or firing 'chaff' to create electronic decoys in the form of false radar echoes. Nor is visible contact necessary to locate the enemy or to launch attacks, for which they also carry a wide variety of sensors. These are all integrated with sophisticated information, command and control systems. The truly revolutionary change has been in the field of submarines. These have changed from being little more than submersibles that could stay underwater for only short periods (twenty-four to forty-eight hours) and which could move only slowly compared to the destroyers that were their principal opponents, to being immensely powerful systems totally independent of the surface and more manœuvrable than anything on it, carrying colossal strike power, missiles and torpedoes and a vast array of high- technology sensors. Similarly, aircraft have grown in striking power and speed, so that naval operations must consider these as major threats.

The US Navy's Beach Jumpers were disbanded in 1972 and absorbed into other units following the adoption of the Line Function Concept, which recognized that all naval vessels could act as potential deception platforms, that the proliferation of radar and electronic warfare specialists meant there was a natural pool of available personnel and that the Beach Jumpers' own equipment was no longer sufficient for this task. In other words, all ships would manipulate their signatures to give off some of the characteristics of high-value targets such as carriers while none would display them all, with the aim of creating target confusion. Since the early 1970s naval stealth and seaborne deception have been dominated by increasingly sophisticated technology. With observation from space and with so many sensors, every aspect of a ship's signature must be considered, and emission control, signature making and the deployment of deception devices (high-tech versions of Cochrane's barrels) now complement attempts to reduce the radar signatures of even very

CHAPTER EIGHT

DECEPTION IN AIR OPERATIONS

'Real life consists of bluffing, of little tactics, of deception, of asking yourself what is the other man going to think I intend to do.'

Jacob Bronowski

THE IMPACT OF aircraft on warfare was immediate and, within a short space of time, dramatic. Steady technical improvements rapidly increased the significance of air power and enabled it in turn to revolutionize operations on land and at sea. In *The Palestine Campaigns* Wavell describes how the devices laid out to mislead the enemy at Beersheba would have been much less effective without the new squadrons and more modern machines from home that enabled the air force to wrest command of the air from the enemy in the late autumn. If they are allowed close to deceptive displays, especially if they are able to fly low enough to make oblique photographs, aircraft are likely to be able to expose them. At the same time air reconnaissance provides a means for the deceiver to feed false information back to the enemy. Of course, anti-aircraft defences are necessary in such cases, in order to make the deception more credible and to prevent the aircraft getting too close to uncover the deception. Before the Normandy invasion in 1944 anti-aircraft gunners in East Anglia were ordered to fire heavily but inaccurately at passing German recce aircraft. Peter Tooley recalled that the gunners 'would go to the pubs in Ipswich at night and of course they were sworn to secrecy. The locals would say to them: "God! We could do better with catapults!" The poor chaps had to grit their teeth and say nothing about it.'[1]

Although capable of defending themselves and of delivering a massive punch when airborne, aircraft are extremely vulnerable on the ground. They are thin-skinned and susceptible to even small-calibre weapons, and their extensive support and technical requirements, including airfields and landing strips, are difficult to conceal and easy to attack. Simple and cost-effective passive measures began with camouflage painting of the aircraft themselves to blend them with the ground when viewed from above, and with the sky when viewed from below. Night-flying aircraft might be painted black underneath, with the paint extending further up the fuselage the lower the plane's normal attack altitude. The appearance late in the Second World War of unpainted American aircraft demonstrated a feeling of security that their bases were safe and that there was no need for camouflage paint.[2]

Decoy aircraft can serve a number of purposes. They can distort the real strength of an airfield or can suggest that a dormant field is active. They can add credibility to decoy airfields or draw enemy fire from real planes, although they would need to be very convincing to fool photo interpreters. The simplest form of decoy is a two-dimensional painting on the ground, but this lacks of the depth and shadows to be effective for long. Carefully designed dummy aircraft may, however, be built from simple materials and if well sited can prove very effective. But the most effective decoys would be genuine non-operational aircraft, always assuming that their non-operability was not too obvious. Major Oliver Thynne showed some ULTRA decrypts to Dudley Clarke after Clarke had been away for several days. Among the reports was one that the Germans could distinguish between real and dummy aircraft by the struts under the wings. When he saw it, Clarke asked abruptly: 'Well, what have you done about it?' 'Done about it, Dudley,' said Thynne, 'what could I do about it?' 'Tell them to put struts under the wings of all the real ones of course!' said Clarke.[3]

Other tell-tale signs that might give away a decoy to a photo interpreter would include a lack of movement, oil stains, tracks or areas of blown grass from engine-testing.[4] These became readily apparent when the first efforts were made to establish a network of decoy sites during the Battle of Britain.

DECOY SITES

Since the First World War many different varieties of paint scheme and netting have been tried in an effort to conceal or counterfeit potential targets or important navigational indicators on the ground. Not only military targets such as airfields, but also important factory sites and communications points such as bridges have been disguised (albeit without great success). The first town to be bombed from the air was Great Yarmouth in Norfolk, which was attacked by Zeppelin airships in 1915. The following year an Essex parson described how

> outside Ipswich is a heath. When the Ipswich folk heard that Zeppelins were coming, they plunged the town into total darkness. Then some went out on the common and lit a bonfire here and there to look (as seen from a great height) like the flare of a big works. They also put up a few acetylene-gas lamps on bushes here and there, to look like lamps at street comers. They lit a lot of squibs to attract attention. Then they went back to Ipswich and waited. Presently, a Zeppelin passed right over Ipswich, hovered over the illuminated common, and dropped a ton-load of bombs about it. Then, having got rid of their cargo, turned and went back seaward, to report to Germany that it had destroyed a great town.[5]

In 1940 Britain was forced to adopt any scheme that might give her even the slightest advantage. Out of desperation a vast web of fake targets was created, on the basis that every German bomb that landed in an empty field instead of damaging airfields or vital industries was a victory. In July 1940 Colonel (later Sir) John Turner, 'a retired officer of drive and initiative' familiarly known as 'Conky Bill', was given the responsibility for building dummy airfields and other deceptions (apart from camouflage).[6] A former Royal Engineer and pre-war Director of Works at the Air Ministry, his first decoys (with which he had experimented before Dunkirk) were crude installations of parallel flares designed to simulate emergency airstrips.[7] These decoy fires (later code-named STARFISH) were operated manually in open countryside a mile or two from airfields, and soon attracted Luftwaffe attention at night. Flare sites in East Anglia and Kent began to report being bombed and a site at North Tuddenham, protecting the airfield at Watton, was hit two nights in a row.[8] Based at Sound City Films, Shepperton, Turner used the full expertise of the film industry's technicians to design and improve these deceptions, 'in order that the enemy should not be depressed at his lack of success'.[9]

By 1 November Turner informed Air Chief Marshal Sir Charles Portal, Chief of the Air Staff, that 'a great deal has been done in the dummy fire line. Twenty-seven have been constructed to guard large Air Force Stations such as depots, training establishments etc.' But now the tactics were modified. Until this point decoy fires had been used as 'secondary' decoys, lit once the target being protected was under attack while any real fires were extinguished as quickly as possible. Turner now attempted to use the fires as a 'direct' decoy with five sites selected around London, to be lit as soon as the defenders of 11 Group RAF knew the Luftwaffe line of attack, in an effort to draw bombing off London itself.[10] RAF personnel were employed to ensure security and in order to give Turner a free hand the organization, now christened 'Colonel Turner's Department', was removed from the normal chain of command.

Originally, the Air Staff had ordered that trenches be dug and filled with oil 'until something better could be evolved'. This proved highly unsatisfactory: the oil fires cracked the soil, the oil leaked into ponds and streams, and a 'chorus of protest at once arose from farmers and others which continued long after this hasty type had been abandoned'.[11] The skill of the film technicians at Shepperton, however, produced three new types of fire: one burning diesel ('full'), one burning paraffin ('medium') and one burning scrap wood ('short'), which could be mixed according to the required effect.[12] Added to these were increasingly sophisticated arrays of lighting ('Q Lights' or 'QLs') in the form of red obstruction lamps, landing Vs and other patterns, and later, fake railway marshalling yards, factories and other key installations. Where possible, these Q-sites were set well away from other important positions on enemy approaches, usually 1,800 to 3,000 yards from a 'parent' airfield, such as Broomfield (covering Maidstone), Lenham (covering Detling) or Upton (covering Ossington). Large air bases at Canterbury and Chelmsford were protected by two sites each and a major installation near Plymouth had no fewer than seven different decoys. There were also a few 'free-standing' targets designed to lure raiders; in a number of cases civilians had to be evacuated from the surrounding area.

However, Turner's success led to further problems. After Luftwaffe pilots reported that they had hit and badly damaged an airfield, their recce planes would expect to photograph this damage shortly afterwards; the inability to do so would risk the security and effectiveness of the decoys. Therefore, fake damage had to be created on the real airfields. Rubble and wreckage, including damaged and unserviceable aircraft, were unfortunately plentiful, though, and could easily be spread around. Dummy bomb craters, which came in two varieties – one for sunny days and one for cloudy weather, with more subdued shadows – were painted on large sheets of canvas and had to be oriented to the sun and turned regularly. After pre-testing at Farnborough, they were nevertheless very effective.[13]

Hitler's Directive No. 17 targeted not just the RAF but the aircraft industry.[14] Factories were therefore equally important subjects for Turner's deceivers. Following an aerial recce, a decoy would be set up a mile or two away and designed to imitate the genuine installation as closely as possible. The blackout regulations would be rigorously enforced at the genuine site, but at the decoy would be subtly less effective. As the Battle of Britain reached its climax, the pressure on the airfields in the south and south-east increased and a new network (K-sites) was planned for daylight hours. Naturally, these sites would have to be very much more complete and convincing, requiring scores of dummy aircraft and prepared with lavish attention to detail. Each site required a runway, maintenance sheds, fuel and bomb dumps and eight or more dummy aircraft. Each 'airfield' was further enhanced with two anti-aircraft machine-guns. As with the Q-sites, these were paired with genuine stations and Marham, Debden and Biggin Hill observe it. The day after it was finished, a solitary RAF plane flew over and circled the field once before dropping a large wooden bomb.[17]

Following the disastrous raid on Coventry on 14 November 1940, it was decided that only large decoy fires lasting several hours could have any chance of drawing the enemy away from the flames of a burning city. The Ministry of Home Security supplied a list of the most important towns and cities and suitable sites were located, so that by the end of December eighteen such sites were in place, mostly in the Midlands. By March 1941, 108 were complete and by July the number had reached 155. By mid-1943, 235 were operational, although by this time the course of the war made many unnecessary and they were beginning to be closed down.[18] The effectiveness of these Starfish fires is debatable. In good weather they had little chance of succeeding and ignition had to coincide precisely with the build-up of enemy air units, but Turner claimed in May 1942 that enemy pilots 'were more easily taken in by the decoys than their predecessors a year ago' and that they 'should be used more boldly'.[19] There were failures, often owing to poor communications, and German prisoners revealed they already knew about some of the sites (which were then moved), but there were many successes. Perhaps the most notable was that on Hayling Island, near Portsmouth, on 17 April 1941. A total of 170 high-explosive bombs, 32 parachute mines and approximately 5,000 incendiaries (ninety per cent of the attack) was dropped on the site.[20] The contribution of the sites was difficult to quantify but, as the official view stated, 'the deception succeeded in materially reducing the weight of bombs falling on towns and cities in the UK'.[21]

The *Blitz* saw the heaviest bombing attacks on Britain, but Hitler's decision to break a cardinal rule of warfare and invade Russia in June 1941 did not mean a complete cessation of bombing by the Luftwaffe, and Turner's men were kept busy. Moonlight reflected on bodies of water was an excellent navigation aid to enemy bombers, as were rivers and canals. In early 1942 another innovation, this time the product of the Admiralty's Department of Miscellaneous Weapons Development, turned waterways into 'roads' by spraying them with a fine layer of coal dust mixed with fuel oil. Attempts to conceal the River Thames were defeated by the scale of the operation, not to mention tides, wind and housewives downstream, who complained bitterly of grimy laundry.[22] But on relatively still water, such as the Coventry Canal, which helped direct aircraft to the heart of Birmingham, it created what looked like an asphalt road even at ground level: one old man and his dog fell in the cut when they innocently went to cross it.[23] In due course, however, electronic navigation aids superseded optical methods, and these required different counter-measures.

HAMBURG AND BERLIN

Camouflage schemes such as this were by no means restricted to the British. Two American reporters, Harry W. Flannery and Howard K. Smith, were in Berlin during 1941. Shortly before Flannery left, he travelled through the

Reich and described how the Nazis had worked hard all summer to make it more difficult to bomb objectives accurately. The most pretentious undertaking was along the East-West Axis, a five-mile long street running from the Brandenburg Gate in the centre of Berlin to Adolf Hitler Platz, which was a guiding arrow to the heart of the capital for attacking airmen. West of the Brandenburg Gate, the Tiergarten lay on each side of the Axis for more than two miles, and here workmen erected steel poles fifteen feet high, stretched wire netting covered with green shrouds of cloth over them, and the tops of pines and other trees hither and thither in rows. Lampposts were covered with green gauze to look like trees and the large Victory Pillar, a monument to the Franco-Prussian War, was covered with netting, with the shiny angel of Triumph that adorned its top dulled so as not to reflect light. Traffic could continue under the netting but the Axis had been blended into the Tiergarten. Elsewhere, the Lützensee, a lake that shone as a guide to aviators, had been covered with a strip of grey netting to resemble a street, although wild ducks remained under the fake buildings. Deutschland Halle and other prominent landmarks in that region were also covered with netting so that they looked like parks with paths running through them. To further the illusion, open spaces in Adolf Hitler Platz and a park near the Axis were filled with pseudo-structures. Smith reported that the first autumn gale ripped great holes in the netting and the whole affair had to be reconstructed.[24]

Hamburg was also camouflaged. To planes overhead the railway station where Flannery arrived resembled a park, and other buildings in the vicinity were blotted from view or given new outlines. More significant were the changes Flannery witnessed to the two lakes in the centre, the Binnen Alster and the Aussen Alster. These lie in the heart of Hamburg's central business district and are formed by the damming of the River Alster about a mile before its confluence with the River Elbe. Both basins and the Lombards river and rail bridge that separates them were easily recognizable from pre-war maps and photographs, and provided a reference point for bombers. The Binnen Alster measured 450 metres by 410 metres – a huge area to camouflage – and the annual snowfall and large chimneys and oil storage tanks made netting impractical. Also, it had to remain open to navigation.[25] Flannery saw that it had been covered with scaffold buildings, with only a narrow lane retained for necessary traffic. To make the air picture more deceptive, the outlines of the basin were reproduced in a part of the larger Aussen Alster, and the bridge that runs between the two was represented in the new position (thus throwing all bomb-aiming calculations off by 300 metres to the north). In the harbour what appeared to be a hilly island with rocks and trees had the end of a factory jutting from the uncompleted park of what, at a few hundred yards, appeared to be a deserted piece of land.[26]

This bold attempt to disguise the landscape itself was ultimately futile. The RAF had seen the whole thing in construction, and during the horrendous 'fire-storm' attacks of July and August 1943 most of it was destroyed. Not all such efforts were fruitless: while it was unlikely that the photo interpreter would be fooled for long, bombers relying on visual identification of the target might be put off just long enough by well-executed camouflage. But when the Soviets painted buildings on Red Square during the summer of 1941 in such an attempt, they found that by autumn the paint was fading and the shadows were all wrong.[27] Other attempts elsewhere to camouflage industrial targets were hampered by their location being well known. Furthermore, paint alone does nothing to deal with shadows, especially on nearby roads. Netting might help, but stereoscopic viewing could still often see through the disguise.

THE SCIENCE OF AERIAL BOMBARDMENT

Before the Second World War the theory of aerial bombardment had struck terror into authorities the world over. Estimates of phenomenal casualties were produced and it was widely held that the bomber would always get through. By 1939, however, the single-seater fighter was faster and more deadly than the bomber, and although the United States Army Air Force made valiant efforts to bomb in daylight, the emphasis shifted to night operations. As a result aerial warfare quickly became a platform for scientists to develop aids carried for navigation and, later, target acquisition; these aids in turn became targets for deceivers. In the summer of 1940 measures were introduced first to jam and then to 'bend' radio navigation beams used by Luftwaffe bombers; later counter-measures were brought in to jam the various radars used to detect the planes.

The first navigational bombing aid the Germans used was called KNICKEBEIN ('Crooked Leg') and was a development of the Lorenz blind approach beam, which was in quite widespread use before the war. The Lorenz system projected two adjacent beams up to thirty miles from an airfield; the left-hand one contained Morse dots and the right-hand one dashes, and a steady note was produced where the beams overlapped. The KNICKEBEIN was a development of this that added a cross-beam from an oblique angle to indicate when a bomber was over the target. It was a simple technique and easily interfered with, especially after a Heinkel He-111 that crashed in March 1940 revealed some detailed information.[28]

With Churchill's support an organization was created to institute counter-measures; by October this had been christened No. 80 (Signals) Wing RAF, and was under the command of Group Captain E. B. Addison. As the summer progressed, KNICKEBEIN transmitters were identified and it became apparent that stronger jamming transmitters (ASPIRIN) and more listening stations would be needed. At about the same time it was decided that re-broadcasting the beam from a different point would effectively confuse the enemy, a process called 'meaconing'. However, by late summer it was clear that the Luftwaffe was aware of this interference and had begun to make rapid changes to frequencies and transmission periods.[29]

It was also at this time that the Luftwaffe introduced two new beam systems that were correctly perceived as blind-bombing systems. The X-GERÄT (X-System) was developed by Dr Hans Pendl and used a set of cross-beams to give accurate instructions for the release of the bomb load. Because it transmitted on a different frequency from the KNICKEBEIN, it was necessary to construct different transmitters (BROMIDE) for counter-measures. Construction of these was aided once more by components recovered from a crashed enemy aircraft and interrogation of the crew, and as early as November the British were successfully interfering with the X-GERÄT system.[30] Unfortunately, and in spite of early warnings provided by ULTRA, there were insufficient jammers in place to prevent the attack on Coventry on the 14th, and those that were in place broadcast at 1500Hz, while the bombers were listening on 2000Hz and were able to work through the interference. Despite the lighting of decoy fires, 449 out of 550 aircraft got through and did enormous damage.

An entirely different system (Y-GERÄT) was also identified. This was operated without a cross-beam and was controlled from the ground, proving very unpopular with aircrew. By February 1941 a counter-measure had been designed, involving re-broadcasting the signal from a powerful transmitter at the BBC's Alexandra Palace. The new system had only been in use on a small scale over Britain, presumably in order to test it and, although it had proved itself remarkably accurate, the test period had also proved its downfall. During March it became apparent that the counter-measures were once more proving effective, by causing the bombers to drop short of their target.[31] By April 1941 the forthcoming invasion of Russia seemed to offer relief to a German air force that had suffered its country's only defeat in the war so far. 'At last a proper war', declared the Luftwaffe Chief of Staff, Generalmajor Hans Jeschonnek, a comment that reflected a widely held view among the increasingly frustrated aircrew.[32]

In their turn, the operations that RAF Bomber Command commenced over Germany in 1940 were initially brushed off by the Nazis as inconsequential. By 1941, however, they were sufficiently effective to necessitate the measures described by Flannery and Smith. By the following year the bombing operations were gaining pace but losses were heavy, largely because of German advances with radar. As early as 1936 they had developed a system called FREYA, which was capable of detecting an aircraft at fifty miles. Two years later they produced the WURZBURG system, which operated at 5000Hz, a much higher frequency than anything the British had at the time, and the first system able to provide anti-aircraft gunners with the height of unseen targets. Together, FREYA and WÜRZBURG made a formidable combination. With two further early warning radars, MAMMUT and WASSERMAN, they formed the basis of the HIMMELBETT system, which was set up along the approaches to the Reich. Each radar was assigned a box of airspace and night fighters operating in the same box would be given instructions by the radar controller in order to bring them close enough to make contact with the bombers on the fighters' own LICHTENSTEIN radar.[33]

By 1942 RAF losses were mounting and it was clear they would continue to do so unless counter-measures could be taken, while the Channel Dash highlighted the serious problems with British radar technology. Night-fighter control radars, anti-aircraft director radars and radars controlling searchlights all operated at around 200Hz. On 28 February 1942 the fledgeling British airborne forces made a daring raid on the coast of northern France and snatched an entire WÜRZBURG radar set from its site at Bruneval, near Le Havre. More equipment was captured later in the year after the German defeat at El Alamein, and from these scientists at the Telecommunications Research Establishment (TRE) at Malvern were able to produce jamming devices. The first of these was MOONSHINE, invented by Dr Joan Curran (of 'Laboratory 15' based at Harvard University, which was the US equivalent of TRE). This amplified the signal from FREYA and gave the impression that a single aircraft was in fact many.

The deceptive possibilities were immediately appreciated and a special unit, No. 515 Sqn RAF, was formed in April 1942 equipped with obsolete Defiant fighters fitted with MOONSHINE. During the spring and summer it was very successful at drawing attention away from genuine operations, attracting anything up to 300 enemy fighters to the wrong location. However, the Germans deployed many more FREYAS and many more MOONSHINE-equipped aircraft would be needed. More significantly, it was only effective in protecting tight-knit formations in daylight, whereas Bomber Command was turning almost exclusively to night operations conducted in widely dispersed streams. No. 515 Sqn flew its last MOONSHINE mission in December and then re-equipped with MANDREL, an American jamming device used ahead of RAF night raids and US Eighth Army Air Force daylight raids throughout 1943.[34]

As the offensive intensified, and given the inherent difficulty of night bombing, a new navigation aid became necessary. One had been conceived as early as 1938, but serious work commenced only in June 1940. This was called GEE and relied on three ground transmitters set 100 miles apart. These emitted pulses in given patterns onto a special GEE map of Europe, and gave an accuracy to within 6 miles at a range of 400. The system was much more effective than any of the previously described German systems, at least until the Germans devised a countermeasure. By March 1942 around one third of Bomber Command aircraft was equipped with the system and the decision was taken to commence operations using it. It proved very effective, but the Germans responded quickly and took steps to jam it, beginning in August. By November GEE was useful only in areas outside German control. That it survived this long, given that several aircraft equipped with experimental versions were lost as early as August 1941, was due to a deception wrought by Dr R. V. Jones. By careful modification and through double agents he succeeded in persuading the Germans during the crucial development period before its operational début, that it was little more than a copy of the KNICKEBEIN.[35]

During the summer of 1943 British losses reached such heights (275 bombers in June alone) that permission was finally given to use WINDOW. This was the British code-name for black paper coated on one side with aluminium foil and cut into strips measuring thirty centimetres by one and a half centimetres – half the wavelength of the German WÜRZBURG and LICHTENSTEIN radars. When dropped in bundles of 2,000 at a rate of one bundle per minute, it would 'snow' the tubes of these radars and create the impression of enormous numbers of aircraft. The principle was not new and had been considered by both sides. Metal strips had been dropped by the RAF in an attempt to confuse German antiaircraft gunners in North Africa in 1941, but this proved unsuccessful (probably because the Germans were using *sound* locating systems). However, fear of giving the idea away held both sides back. Tests continued during 1942, but Air Chief Marshal Sir Sholto Douglas, Commander-in-Chief Fighter Command, persuaded his opposite number at Bomber Command, Air Chief Marshal Sir Arthur Harris, not to use it in case the Germans should choose to renew their own bombing offensive. In reality, the Germans were so horrified by the implications from the trials of their own version, code-named DÖPPEL, that Luftwaffe chief Reichsmarschall Hermann Goring ordered all evidence of it to be destroyed.

Other developments finally helped provide an excellent opportunity to use WINDOW. These included the introduction of H2S radar, which gave a picture of the ground, and the directional system OBOE, which was particularly suited to the Mosquito aircraft of the Pathfinder force. Furthermore, the American SCR 270 night-fighter radar – which could distinguish between aircraft, WINDOW and the SERRATE radar detection device – had also recently been introduced. (The latter enabled British fighters to home in on the German night fighters' LICHTENSTEIN radar.) Vast quantities of WINDOW would be needed (around 400 tons per month) and when suitably large stockpiles had been built up it was carried for the first time in the GOMORRAH

Chapter 1. The French surrender at Goodwick Sands, Fishguard, 1797. Among the watching crowds are women in traditional costume, who may have contributed to the French decision to capitulate

Chapter 2. Quaker guns near Centerville, 1862. 'Prince John' Magruder had just fifteen real guns on the York-James Peninsula, but by supplementing them with Quaker guns he stopped an army

3. Before Allenby's climactic Battle of Meggido in 1918 thousands of dummy horses were made from canvas and wood. Sleighs drawn by mules raised clouds of dust when they would be expected to go to water

4. A store at the Royal Engineers Camouflage School during the First World War, containing hundreds of figures, papier mâché heads and other deceptive devices

" Of course, at the moment it's still just a suspicion."

5. During the Second World War, fear of 'fifth columnists'
led to some absurd stories, satirized by this Osbert Lancaster
cartoon. But the panic caused in rear areas could be genuine

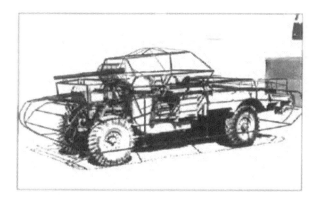

6. Dummy tanks in various stages of construction from wood and canvas. Such dummies have often been improvised using whatever materials happen to be available

7. Dummy tanks on the move in the Middle
East during the Second World War

8. More elaborate materials are sometimes required
for camouflage, such as this plaster 'brickwork' being
made to cover pillboxes in France in 1940

9. A decoy built by Polish troops in Italy during the
Second World War

10. An inflatable 'Sherman tank' – quick and simple to use.
But such equipment has drawbacks: it will sag or expand with
atmospheric changes, and rubber tanks leave no tracks

11. Sometimes entire defences are built on a lie.
Here, 'Hurricane' fighters are assembled at a factory
in Cyprus during the Second World War

12. A 'Sunshield' device for turning a tank into
a 'truck'. Resources that cannot be hidden can
at least be disguised as something else

13. The dummy railhead at Misheifa, 1941. The entire marshalling yard was built from wood and canvas, and rails were made from four-gallon tins normally used for carrying fuel and water

14. Snipers in training. In the top photograph three snipers are aiming at the camera, showing how difficult it is to see properly concealed troops even at short range

15. The *Admiral Graf Spec* burns in Montevideo harbour, 1939. Clever diplomatic noises forced her captain to scuttle her rather than face an enemy that she could have crushed

16. These R-class battleships and aircraft-carrier
in Scapa Flow in July 1940 are in fact fleet
tenders disguised with canvas and wood

17. *(above)* HMS *Hesperus* in a typical dazzle camouflage scheme and *(below)* HMS *Belfast* showing false bow
and stern waves, all designed to hinder targeting

18. Airfields are conspicuous and vulnerable targets, and
efforts to make them harder to identify are essential

19. A dummy airfield is an enormous undertaking if it is to convince the enemy. It must possess all the characteristics of the real thing, including many buildings

20. Among the requirements for successful deception is
the ability to recreate damage. While cinematic special
effects techniques are desirable, they are not always available, as this crude decoy fire equipment shows

21. This runway at Rheine in north Germany appears to
be inoperable. But examination through a stereoscope
of this photo and another one slightly offset revealed
that the bomb craters on the runway were fakes

22. Hamburg before and after camouflage, 1941. A fake Lombards bridge has been built some distance to the north of the real one and the Binnen Alster covered over

23. A cloud of Window released over Essen in March 1945. Often referred to today as 'chaff', it is used by ships and planes as a decoy against radar-guided missiles. Flares are used to decoy heat-seeking missiles.

24. *(above left)* Dummy landing craft displayed in an estuary and *(left)* a dummy 25-pounder and Quad tractor in southern England before D-Day

25. Smoke-generating equipment deployed for the crossing of the Rhine in March 1945. Huge smokescreens can be made to last for days on end

26. Members of a counter-gang prepare to go on patrol in Kenya during the Mau Mau rebellion. The man on the right is a white agent disguised as a rebel guerrilla

27. As part of Operation Desert Storm, this Psyops leaflet was carefully designed to reinforce the Iraqi fear of an amphibious assault on Kuwait. Such leaflets were put in bottles and began floating ashore one day before the Allied deadline of 15 January 1991

28. An Ml Abrams decoy. Modern decoys must cover all spectra and this example comes with a heating element and a 1kW gas generator

29. A Saab Viggen fighter sits next to a decoy. An enemy pilot flying at high speed would find it almost impossible to distinguish the real one

The RAF now took to launching diversionary attacks by small groups of Mosquitoes dropping large quantities of WINDOW away from the genuine direction of attack. It was found that dropping WINDOW in controlled patterns could produce the image of an aircraft formation on a radar screen. In due course, however, anti-WINDOW modifications were developed for German fighter control and gunlaying radars, together with homing devices for H2S, and loss rates among the bombers soon began to climb again. The next RAF move came on the night of 26 October 1943, when 569 bombers were sent to raid Kassel. It had been found that not only directional beams, but also voice transmissions, could be broadcast into the heart of Germany. Fluent German speakers could therefore send 'spoof' instructions to the German night fighters. Once more the Germans found a counter-measure and thereafter the British resorted to long 'test' transmissions to take up 'airtime'. When this method failed, they gave readings from particularly boring German classical works or recordings of Hitler's speeches.[38]

In August the Germans had introduced the WILDE SAU ('wild sow') tactic, whereby single-seat fighters would be concentrated over the bombers' target and use searchlight illumination for target acquisition. Unfortunately for the Luftwaffe, this meant that its controllers were even more reliant on accurate information as to the bomber force's target and direction of approach. The short-range single-seaters' lack of radar gave them little other chance of intercepting the bomber stream, and if they fell for a diversionary raid they would end up chasing Mosquitoes. The British quickly caught on to the WILDE SAU and launched more diversionary missions, with their deception tactics becoming ever more complicated.[39] In spite of this chicanery, it was the Luftwaffe that held the upper hand during the period from November 1943 to June 1944, as a result of improved radar and mass attacks by every available fighter.

Harris was convinced that a final effort by Bomber Command would end the war, and he chose to make it in what became known as the Battle of Berlin. At the same time the Luftwaffe introduced a tactic called ZAHME SAU ('tame sow') which fed radar-equipped night fighters directly into the bomber stream, and began intercepting bombers as far out as the North Sea. It proved a serious defeat for Bomber Command, although one that Harris did not acknowledge until March 1944, by which time 1,128 aircraft had been lost. The culmination was the disastrous Nuremberg raid of 30 March. Careful German monitoring of H2S transmissions showed two Mosquito feints on Kassel and Cologne for what they were, and allowed 246 fighters to attack the main force as it was monitored along its long route into southern Germany. The loss of 107 bombers was catastrophic. However, the Battle of Berlin also coincided with the formation on 23 November 1943 of 100 (Bomber Support) Group RAF, under Air Commodore (later Air Vice-Marshal) E. B. Addison – previously CO of No. 80 (Signals) Wing, which it incorporated

– controlling the various electronic warfare squadrons and several Mosquito fighter squadrons transferred from Fighter Command. It was 100 Group that would ultimately give Bomber Command the edge.

It took some months to train and bring 100 Group up to strength but, despite the disaster of the Nuremberg raid, once full-scale operations were initiated main force losses were cut by eighty per cent.[40] Four days before Nuremberg, a raid was launched that appeared to be following a route leading deep into Germany. Instead, it swung away without warning to bomb Essen and return to Britain having lost only nine bombers (at least three from anti-aircraft fire); sixteen German fighters were destroyed in the raid, many in accidents. Following operations in support of the Allied invasion of Normandy, Bomber Command resumed its full-scale offensive against the Reich in June 1944. The task of 100 Group was aggressive as well as deceptive. Mosquito squadrons were now also equipped with an improved SERRATE system and a device called PERFECTOS which was designed to trip the transponders carried by the night fighters in the same way as their ground stations.

These combined measures made defence almost impossible and create *moskitopanik* among the Germans, who found it increasingly difficult to identify and intercept the bomber streams while their night fighters were increasingly vulnerable to those of 100 Group, particularly around their own airfields or flying holding patterns while awaiting orders. The ace Hans

Krause resorted to approaching his base in a power dive from 10,000 feet, followed by an immediate landing on the dimly lit airfield to throw any lurking British aircraft off the scent. Naturally, such extreme measures led to a sharp increase in losses through accidents and contributed further to the war of attrition that eventually led to the defeat of the Luftwaffe.[41]

Other squadrons within the group operated Flying Fortress and Liberator aircraft, all equipped with a variety of electronic equipment, such as AIRBORNE CIGAR and MONICA, and carrying WINDOW and marker flares to lay false trails. The German air and civil defences were increasingly stretched as the bombers ranged further and, thanks to 100 Group, received less attention. On the night of 4 December 1944 Bomber Command launched four full- scale raids, involving 892 heavy bombers, on Karlsruhe and Heilbronn in southern Germany and Hamm and Hagen in the Ruhr. A total of 112 aircraft from 100 Group flew in support over 'Happy Valley', as the central Ruhr was known. Almost a hundred night fighters were sent to counter the false raid and kept occupied there until it was too late for their deployment against the bombers, which suffered only one and a half per cent losses while 100 Group inflicted substantial losses on the enemy. The following night 553 aircraft attacked Soest, Duisburg, Nuremberg and Mannheim, supported by 76 aircraft from 100 Group; only one aircraft was lost. The following night no fewer than 1,291 bombers attacked Bergesburg, Osnabrück, Giesen, Berlin, Schwerte and Hannau, supported by 89 aircraft from 100 Group, of which one was lost. Loss among the bombers was 1.7 per cent.[42]

Towards the end of the war the Germans continued to develop counter-measures. Radars could be jammed, although at this time jamming techniques were primitive. One very imaginative idea was the use of radar reflectors, particularly floating 'corner' reflectors which, when moored in large numbers around water bodies, could completely alter the representation of these bodies on Allied radar sets. They could also be used to simulate targets in an attempt to induce bombers to drop their bombs in a nearby lake. These reflectors were almost useless in clear visibility, but in bad weather or at night they could be very effective and were easily relocatable. Had they been used in large quantities together with effective jamming, they might have proved very effective but the Germans either hit on the idea too late or lacked the materials or failed to realize its potential.[43]

Although it played a significant part in the success of Bomber Command's operations and saved countless aircrew lives, the deception wrought by 100 Group's technology was not in itself decisive. Probably the single greatest factor in the ultimate defeat of the Luftwaffe was the ability of long-range fighters to support American daylight raids and heavy attacks on fuel plants. The P-51 Mustang was one of the outstanding aircraft of the war, and on 11 April 1944 it provided an example of one of the simplest bluffs in aerial warfare. The citation for the Distinguished Service Cross sites eventually discovered mysterious curved ramps at the Bois Carré in northern France. The Allied air forces dropped some 30,000 tons of bombs on various related targets (more than half as much again as the Germans had dropped on London during the *Blitz*) and completely disrupted the programme, forcing the Germans to start again from scratch. This time they simplified the bases and carefully concealed them, mostly in wooded areas between Le Havre and Calais. Everything was kept under cover and security was severely tightened. When the first of these new sites was discovered by the patient photo interpreters in April 1944, with the Normandy invasion barely weeks away, the Allies felt they once more had the upper hand. But the Germans adopted a very subtle deception by secretly reoccupying many of the original sites, by now abandoned and ignored, carefully leaving all the wreckage undisturbed – a sort of reverse decoy – which the photo interpreters simply could not see.

The V-bomb onslaught commenced on 13 June, and on the night of 15/16 June 217 missiles were fired. The Twenty Committee immediately began feeding false information via agents GARBO, BRUTUS and TATE suggesting that they were falling in a wide arc to the north and west of the city centre. Two days later GARBO wrote a letter decrying the effectiveness of the weapon and stating that only seventeen per cent of them had landed in 'Greater London'. The true figure was nearer thirty per cent and as more missiles fell the Air Ministry was able to assess the mean point of impact (MPI) as being around North Dulwich station.[46] The aim of the Air Ministry and the Home Defence Executive was therefore to prevent the Germans from correcting their aim, as Dr. R. V. Jones described it, by giving

> correct points of impact for bombs that tended to have a longer range than usual but couple these with the times of bombs that had in fact fallen short.... Therefore if they made any correction at all, it would be to reduce the average range.[47]

Extensive discussions followed among the various government departments affected by the proposed deception, some of whom were concerned at any attempt to divert the bombs since they might cause more harm. At the same time the demands of the major Normandy deception plan, FORTITUDE SOUTH, had also to be considered and GARBO was 'arrested'. As as a result, on 12 July both GARBO and BRUTUS were prohibited by the Germans from sending information regarding the flying bombs. However, by the time a policy had been agreed by the War Cabinet, the Allies were close to overrunning the launch sites in northern France. What served most to deceive the Germans was in fact reports from a freelance agent in Madrid called OSTRO, whose accounts were based on gossip, imagination and whatever he could glean from the newspapers. The real success of CROSSBOW from the British point of view was that it maintained the credibility of its agents. This proved invaluable when the V-2 rockets began falling later in the year. TATE and ROVER put over information on the basis previously described for the V-1s, and although there is no clear evidence that it was specifically these reports that led to a shift in the MPI, a report from the Ministry of Home Security indicated that a further 1,300 people would have died and 10,000 been injured, as well as severe disruption caused, had the missiles indeed been falling between Westminster and the docks, as the Germans were led to believe.[48]

AERIAL DECEPTION SINCE THE SECOND WORLD WAR

Tactical deception and developments in electronic warfare made a considerable contribution to Allied success in the Second World War and saved countless aircrew lives. Perhaps more significantly, air units were formed for the first time and equipped specifically to carry out this role. As is so often the case, these lessons were quickly forgotten in the immediate aftermath of the war, but were then revived and developed, and since the war there have been phenomenal developments in both aircraft and sophisticated avionics systems. Aircraft themselves are faster and more powerful, ranging from fast agile fighters through versatile multi-role fighter-bombers to the colossal B-52 bomber, which dwarfs the Lancaster and Super Fortress of the Second World War. More significantly, aerial warfare no longer involves getting as close as possible to one's target in order to shoot directly at it with cannon or machine-guns, or to lob free-fall bombs. Instead, it relies on sophisticated missile technology, based on radar or infra-red systems. Similar systems now equip ground-based air defence units and tactics have been transformed, leading in turn to defensive flares and chaff (bundles of WINDOW) to act as decoys and draw off attacking missiles. Flying at low altitude and high speeds may put aircraft under effective radar cover, but also makes them vulnerable to ground fire.

The development of 'Wild Weasel' electronic counter-measure aircraft led to specially equipped fighter-bomber units, designed to seek out and destroy enemy radar-controlled gun and missile systems in a role called Suppression of Enemy Air Defence (SEAD). If an enemy radar is activated, it can be located and attacked with anti-radiation missiles that home in on radar emissions. Since the Vietnam War such units have formed a crucial element in securing control of the air. More recently, so-called 'stealth' technology has been designed to make aircraft radar 'invisible', using radar-absorbent materials, carefully designed shapes and 'cancellation' technology (avionics that predict an aircraft's reflective signal at a given frequency and angle in order to transmit a signal which will cancel the signature). 'Stealth' technology also includes techniques to mask sound and infra-red sources such as engine exhausts. Although the primary threat was seen for a long time as being from radar-guided weapons, a US Department of Defense study showed that in the ten years to 1985 ninety per cent of tactical aircraft destroyed were victims of infra-red systems.[49] Such technology is increasingly being applied to land and naval units while towed decoys, in use at sea since the 1940s as a method of countering acoustic torpedoes, have been adopted since the 1980s in the form of a small radar jammer towed behind an aircraft. However, counter-'stealth' techniques were soon identified, since it seems each measure is quickly met by a counter-measure in a process that seems likely to continue for many years to come.

or so British and Canadian troops who had been preparing for years, one and a half million Americans were
in Britain – some of them veterans of North Africa and
Italy, but many more newly arrived from the USA, after
a six-month training period, having never heard a shot
fired in anger. Millions more waited to follow direct from
America, and naturally there was considerable nervousness at all levels of command at the enormous risks such
an operation entailed. These fears were apparently justified on 28 April 1944, when convoy T4, taking part in the
pre-invasion Exercise Tiger in Lyme Bay was attacked by
marauding German E-Boats and 'trapped and hemmed
in like a bunch of wolves circling a wounded dog'.[1] Seven
hundred and forty-nine American servicemen lost their
lives, and the vulnerability of projecting a vast force across
defended seas was painfully demonstrated. As Patton once
said: 'In landing operations, retreat is impossible.'[2]

On the far side of the Channel the Germans were
expecting an invasion. It was understood by OKW that
the invasion represented a decisive point in the conduct of
the war. If it could be successfully defeated than it would
release as many as fifty divisions for the Eastern Front. The
Grand Alliance would be paralysed if not shattered, and
such a defeat would also provide the time the Germans
needed to deploy the new weapons – jet aircraft, long-
range submarines and V-weapons – that Hitler believed
would ensure final victory. But defeating the invasion
was complicated by the difficulty of predicting when
and where it would occur and Hitler was obsessed with
defending every inch of the European coastline, however

unlikely the threat. The landings could come, he said, 'on any sector of the long western front from Norway to the Bay of Biscay, or on the Mediterranean – either the south coast of France, the Italian coast, or the Balkans'.[3] This meant spreading his resources thinly and hoping to maintain strong enough reserves to throw the Allies back once they had shown their hand. In fact, whatever the Allies might have wanted to do, the fundamental considerations of supply and air cover presented them with only two feasible alternatives for a large-scale invasion, Normandy and the Pas de Calais.[4] Either way, it was apparent to both sides that the Allies would have only one chance to effect a lodgement and, as Hitler himself said, 'once the landing has been defeated it will under no circumstances be repeated'.[5] The Allies could not afford to fail and would need every possible assistance.

The defences in northern France against a seaborne assault were formidable. Under von Rundstedt, Germany's senior field commander, were two Heeresgruppen ('army groups') in France and the Low Countries. Although von Rundstedt was under no illusions as to his true power under Hitler's control ('my sole prerogative', he said later, 'was to change the guard in front of my gate'),[6] he was known to dispose of some sixty divisions and for all Hitler's interference he would still be a prime target for the deceivers. Heeresgruppe B comprised Fifteenth Army, which covered the area from the Scheldt estuary in Belgium westward, including the Pas de Calais almost to the River Orne. On its left flank Seventh Army covered Normandy and Brittany. In the French interior and south Heeresgruppe G comprised First and Nineteenth Armies. Elsewhere, there were strong forces holding Italy, Scandinavia and the Balkans. The diversion of these strong enemy forces away from the chosen crucible of operations in Normandy, both from outside and within the Western theatre, was the aim of the carefully co-ordinated deception plans.

GERMAN INTELLIGENCE

Allied intelligence agencies were well aware of the state of affairs within the Reich and occupied territories. The assistance of resistance organizations across Europe and the almost total control of the skies for the purpose of aerial reconnaissance gave them a huge advantage in information gathering, to say nothing of the insights provided by ULTRA. Furthermore, various German agencies were dissipating their energies on the creation of a *feindbild* or picture of the enemy and his intentions. FHW was but one agency dealing with intelligence, and von Roenne found himself increasingly competing not so much with the Abwehr as with the Sicherheitsdienst (SD, the security branch of the SS). When the Abwehr was abruptly disbanded in February 1944 as a result of its involvement in anti-Nazi intrigues, German intelligence was thrown into confusion from which both FHW and another rival, Amt VI of the Reichssicherheitshauptamt (RSHA),* com-

---

* The SS Reichssicherheitshauptamt or SS Hauptamt, the Reich security main office, was a powerful bureau of Heinrich Himmler's SS

manded by Brigadeführer Walter Schellenberg, found themselves reporting to Ernst Kaltenbrunner's SD. Kaltenbrunner's tendency for personal reasons to send what he regarded as 'vital' information straight to Hitler or his lackey Generaloberst Alfred Jodl, chief of the Wehrmachtführungsstab, meant little intelligence worthy of the name was properly circulated.[7] These agencies, already reeling under a welter of claims and counter-claims from false radio intercepts, double agents, camouflage, decoys and diplomatic rumours, found themselves unable to filter truth from fantasy, and the Allies knew it.

THE OVERALL PLAN

At the strategic level, it was hoped that by threatening widely scattered areas of occupied Europe the Germans would be persuaded to maintain unnecessary strength in these peripheries, thus leaving France relatively weakly defended.[8] In October 1943 Bevan put forward a draft plan code-named JAEL. This would threaten major operations both from Britain and in the Mediterranean, but at this stage none of the ideas for the actual landings had been

---

headquarters set up on 27 September 1939 by Reinhard Heydrich. Left over from the SD were personnel and administrative branches (Amt I and Amt II) and newly created were Amt III (Interior), Amt IV (Gestapo), Amt V (Criminal) and Amt VI (Foreign Intelligence). Bureaucratic, over-compartmentalized and inefficient, it nevertheless served Heydrich well in his climb to power until his assassination in 1942. Himmler personally headed the RSHA until Kaltenbrunner took it over in January 1943.

formalized and the deception plans were correspondingly unconvincing.

At the same time the planning staff at COSSAC, putting together the details of the genuine operation, produced a tactical deception plan named TORRENT that alarmed Bevan. He understood the need to tie deception to operational reality and, among other things, disliked the amateurish way that it planned for the period after D-Day. He persuaded the Combined Chiefs of Staff to shelve it until after the meetings between Roosevelt and Churchill ended in Cairo in January 1944, after which the final details were to be nailed in place and TORRENT would then form the basis of 'the OVERLORD cover plan'.[9] It was, however, agreed that this plan should be provisionally circulated immediately, as some measures would take time to become effective. Radio silences would need to be imposed so that the enemy would not recognize the genuine silence masking OVERLORD. Dummy build-ups and genuine concealment would also require time to implement.[10]

The result of the various conferences was the decision to launch Overlord in May 1944, and it was also agreed to launch Operation ANVIL simultaneously against the south of France. Eisenhower was nominated Supreme Allied Commander at Tehran* in December 1943, and JAEL was renamed Operation BODYGUARD before receiv-

---

* It was here that Churchill made the famous comment that heads the chapter (W. S. Churchill, *The Second World War,* vol. V, *The Hinge of Fate,* p.338).

ing its final approval from the Combined Chiefs of Staff on 25 January 1944.[11] Implementation could now begin in earnest, and Bevan had Noël Wild from 'A' Force appointed to head Ops 'B', the critical executive post for deception within Eisenhower's Supreme Headquarters Allied Expeditionary Force (SHAEF). The aim would be to persuade the Germans that although a large-scale cross-Channel operation was indeed being prepared, it was by no means the only planned stroke and that anyway it was not due until late in the summer. Brigadier-General Harold R. Bull of the G-3 (Operations) staff at SHAEF, described the aims of the cover plan as being to:

> Cause the Wehrmacht to make faulty troop dispositions by military threats against Norway ... Deceive the enemy as to the correct target date and target area of Operation NEPTUNE ... Induce the enemy to make faulty tactical dispositions during and after NEPTUNE, by threats against the Pas de Calais.[12]

Within BODYGUARD the principal plans consisted of the threat to Scandinavia, which would be developed through FORTITUDE NORTH, supported by diplomatic deception in the form of GRAFFHAM and ROYAL FLUSH. The threat against the Pas de Calais would be developed as FORTITUDE SOUTH, supported by IRONSIDE. Meanwhile in the Mediterranean elements of ROYAL FLUSH would further support operations ZEPPELIN and VENDETTA.

Strategic Deception: Fortitude North, Zeppelin and Vendetta

The Germans knew that they could not weaken the Eastern Front and to begin with believed that only France would be targeted for invasion. OKW held that a force of eight divisions could be assembled from the periphery of occupied Europe – Scandinavia and the Balkans – to drive the invasion back into the sea.[13] By January 1944 that synopsis had to be altered as threats appeared to indicate that, although the main invasion would cross the narrow Straits of Dover, subsidiary attacks against the very areas from which their reaction force was to come could also be expected, and thus these areas could not afford to be weakened.

After many years' experience the deception staff knew that in order to ensure the coherence of the myriad tiny details necessary to build a credible picture in the minds of their targets, their cover plan should be produced as if it were for a genuine operation. For the benefit of NEPTUNE, the best time to launch an invasion of Norway would be after D-Day, so that all the formations in Norway would remain there while the genuine operation took place. Scotland was the obvious assembly area for any genuine invasion of Norway, so it was also the obvious assembly area for any fictional invasion of Norway. A detailed plan involving an initial landing of two divisions in the north to capture Narvik and designed to attract German forces to meet it would be followed by a landing to capture Stavanger airfield and land further reinforcements, with Bergen, Trondheim, and eventually Oslo for objectives.

As well as persuading the Germans to retain forces away from the crucial battlefields, a principal aim of FORTITUDE NORTH was to make them believe that the invasion of France was planned for a later date than in reality. The scenario therefore suggested an attack timed one month before the invasion of France, which would supposedly be at the end of July. Once this was established, the notional British Fourth Army could be created in order to carry out the notional plan.

Fourth Army was based on the existing Scottish Command under Lieutenant-General Sir Andrew Thorne. Usefully, 'Bulgy' Thorne had been the military attaché to Berlin in 1934 and was known personally to Hitler. Although Thorne commanded the formation, the man appointed to run the detailed deception was fifty-year-old Colonel Roderick MacLeod, who, the LCS discovered, was an authority on military legerdemain. Many of the troops forming the deception were genuine enough, including the 3rd Infantry Division, which, while carrying out tough assault training on the Moray Firth for the real invasion, was blissfully unaware that this served a dual purpose. Percy Nunn, a driver with 52nd (Lowland) Division, recalled receiving lessons in Norwegian and hard training in mountain warfare. Long after the war he remained doubtful that it was all merely a sham. In December 1943 the division was moved north and

> We were issued with shoulder flashes which said 'Mountain'. They were quite obvious. We all sewed those on and that was added to our divisional sign,

so that we were a pukka Mountain Division. In Dundee, down in the harbour area, you couldn't miss a considerable display of snow-clearing equipment, giant snow-ploughs and bulldozers. It looked like business.[14]

The real and notional units came to a combined total of two corps of six divisions.

MacLeod's greatest contribution came in the form of providing the detailed radio traffic that a real army would produce, under the code-name SKYE. Initially he had serious difficulty getting enough quality staff and proposed that he restrict his efforts to simulating a corps. But SHAEF insisted that an army would be needed to assault Scandinavia, so an army there would be.[15] Since Scottish Command had no ready-made specialist signals unit, a team was made up from II Corps and 9th Armoured Division, both formations in the process of disbandment, to which were added personnel from Scottish Command and various units.[16] To be convincing there needed to be not only the main headquarters elements and large formations such as notional 80th Division, but the myriad supporting units necessary to administer so large a force, thus stretching MacLeod's resources and ingenuity. Consequently, 55th Field Dressing Station, 87th Field Cash Office and a film and photographic section were among the units invented to give authenticity to Fourth Army.[17] The radio traffic dealt with ski training, requisitions for heavy boots and the performance of tank engines in sub-zero temperatures, all details appropriate to an invasion of Norway.[18] Notional naval forces were created and notional amphibious training carried out between the various elements.

The radio deceivers expected an audience. Germany's various wireless intercept and cryptanalytic services were skilled at intercepting, locating and interpreting messages and patterns. They could absolutely not afford to ignore the traffic around Britain. Ensuring that they believed what they heard was another matter, however, and enormous care was needed in preparing the transmissions since a single mistake could blow the whole deception.[19] Similarly, the Germans could be expected to seek corroboration wherever possible, and various visual deceptions were prepared for the Luftwaffe to photograph. Every available ship was anchored in the Firth of Forth, together with makeshift troop barges and landing craft, often lovingly embellished with smoke from funnels, lines of laundry, and music blaring from cabins. A carrier-borne aerial reconnaissance of the Narvik area, Operation VERITAS, was carried out on 26 April. The RAF simulated the transfer of four medium bomber squadrons from East Anglia to eastern Scotland and made displays of real and dummy aircraft around Peterhead, Glasgow and Edinburgh. Apart from twin-engine 'bombers', inflatable rubber tanks were 'driven' into positions complete with 'tracks' and not quite effective camouflage. Plywood gliders, piles of crates and fuel drums were accumulated. Local newspapers joined in, carrying spurious reports of sporting and social events, including marriages of soldiers and local girls.

In Iceland the double agents COBWEB and BEETLE reported a build-up there as aimed at northern Norway and on 12 April BRUTUS described the insignia of Fourth Army and its fictitious components, II Corps and VII Corps, to his eager controller. He also mentioned an American corps that he could not identify and spoke as though the Germans were totally familiar with the operation and were merely unsure as to its exact date. On 4 May he reported the arrival of the Russian Klementi Budyenny, ostensibly to co-ordinate a Soviet attack from Petsamo. Meanwhile, FREAK 'met' an American liaison officer from XV Corps in Northern Ireland attached to Fourth Army in Edinburgh, for which he received congratulations from his control. HAMLET sent a message from his 'contact' in the Ministry of Economic Warfare, saying that the ministry had removed files under strictest secrecy referring to installations in Alesund, Bergen, Oslo, Narvik, Stavanger, Moss, Porsgrund, Kristiansand and Trondheim, a beautifully subtle way of telling the Germans that the invasion would be divided in two, as per the cover plan.[20] MUTT, JEFF and in particular GARBO were enormously valuable, sending a steady stream of information confirming the build-up in Scotland.

Supplementing FORTITUDE NORTH was Operation GRAFFHAM. This aimed to persuade the Germans that the Swedes were being enlisted to assist in the proposed attack on northern Norway It was a combined effort by the British, Americans and Soviets, involving various diplomatic subterfuges and other elements, including the rigging of the Stockholm stock market to raise the price of

Norwegian securities and imply the anticipated liberation of Norway[21] GRAFFHAM produced no dramatic results, but it did have some effect. Most notable was the visit of 'Air Vice- Marshal' Thornton, former Air Attaché, to his old friend General Nordenskiold, the Commander-in-Chief of the Swedish Air Force. Careful to arrive and leave by the back door in civilian clothes (since he knew the Germans would photograph him), he proposed that the planned invasion of Norway would lead to many civilian deaths and that Nordenskiold should seek to persuade the Swedish government to enter Norway as a police force to protect the rights of non-combatants.

The risks run by deceivers are amply illustrated by the fact that Thornton, in reality an air commodore, had difficulty with the Air Ministry Pay Branch over his expenses claims at the air vice-marshal rate until a member of the air staff came to his rescue.[22] The effect of all this was partially successful. FHW duly reported on 28 May that

> Credible-sounding reported overtures by an English Air Force officer in Sweden which were apparently aimed at obtaining air bases in Sweden for invasion purposes, may be interpreted as a hint of a minor landing operation being planned in south Norway or Denmark. The likelihood of a more powerful landing in those parts as part of a large operational strategy is still considered to be slight.[23]

The Germans believed in the existence of Fourth Army, but did not take the threat as seriously as was hoped, given the problems that providing air cover for such a scheme would involve. FHW observed as early as 1 March that

What the enemy leadership is up to in the present stage of operations is to do everything possible to tie down the German forces on subsidiary fronts, and indeed divert them from the decisive Atlantic front; and seeing that they have already tried to do this in Italy, it seems possible that they have decided to do the same in the Scandinavian region.[24]

Indeed, those realists noted on 25 May that shipping in east Scottish harbours was rather sparse and that it had considerably increased in the Portland area, 'permitting the transport of ten or more divisions'.[25]

While FHW had few doubts as to the true *schwerpunkt*, some eighteen divisions – around 200,000 men – were being maintained in Scandinavia, well away from the decisive front. These forces were put in a state of readiness in May, restricting the depletion of their garrisons, and in the middle of May a first-class division was sent to Norway Although OKW decided that the troops in Scotland were capable of only limited operations against Norway and Denmark, they believed that the possibility of an attack remained and the Führer refused to weaken what he termed his 'Zone of Destiny'.[26] This would doubtless have amused the twenty-eight officers and 334 other ranks who finally served as Fourth Army.

Ironically, as von Roenne at FHW came to see the threat to Norway for what it was, so this reinforced the conviction that if the Allies were prepared to go to such lengths, it could only be to divert attention from the most obvious solution, a landing in the Pas de Calais. The Allied deceivers thus inadvertently succeeded in luring the Germans into the trap of a double bluff. With every piece of evidence, real or illusory, they would now reinforce their determined belief in the threat to the Pas de Calais. Since this area was already being defended by three times as many troops as any comparable stretch of coastline, it seemed obvious to many Germans that the Allies would do everything in their power to persuade them that the landings would come elsewhere; and the German response was that with the Allies' notoriety for skilful deception the best course would be to keep their troops where they were and remain alert.[27]

Meanwhile, other plans were instituted in the diplomatic field, notably ROYAL FLUSH, designed to bring pressure on neutrals around the periphery, in particular Spain, Turkey and Sweden. However, the German intelligence agencies were singularly unimpressed by information derived from Allied embassies in neutral countries. FHW reported that 'for quite a long time the Anglo-Saxon embassies have been recognized as outspoken deception centres, from which not a single piece of reliable news, but many false reports, have been coming.'[28] More significant was the Mediterranean deception plan, Operation ZEPPELIN, no less comprehensive than FORTITUDE, and utilizing the considerable talents of 'A' Force, among others. The aim was to mount a threat against the Balkans and thus prevent the transfer of troops from the Mediterranean theatre to France. Clarke employed every trick in the deceiver's armoury, using every branch of the armed forces and including radio, visual and propaganda exercises involving Ninth, Tenth and Twelfth Armies, once more comprising real formations (10th Indian Division) and notional ones (8th Armoured Division, 34th British Infantry Division). Hitler referred to 'the proven presence of battle-strength enemy divisions in Egypt'.[29] In reality, there were just three divisions, none of which was capable of taking the field. The German High Command was receptive to these threats and a directive of 1 January 1944 from OKW to the Commander-in-Chief South-East even went to far as to warn him that Allied stress on north-west Europe might be a cover for a blow in the Balkans.[30]

Operation FOYNES was a plan initiated by the LCS to conceal the transfer of eight battle-trained divisions from Italy to Britain which began in November 1943. The aim was both to exaggerate the strength in the Mediterranean and to minimize the build-up in Britain as part of the policy of 'postponement'. The sudden appearance of Africa Star medal ribbons on battledress throughout Britain was put down to 'instructors' brought to train the less experienced formations due for the invasion. Before the end of 1943 the Germans were already beginning to notice that these eight divisions had disappeared from the Italian front and might have returned to Britain. A letter from a British soldier that fell into German hands told them that 'the Scottish division which came from Italy will presumably take place in the great events due to take place.' They therefore deduced that 51st (Highland) Division had returned to Britain and was to take part in the invasion. However, in spite of this and other indications, including the transfer of landing craft, they tended to place the FOYNES divisions in the base areas of the Italian front and did not accept the transfer of these divisions as fact until notified by one of the controlled agents.[31]

By March the Germans were beginning to consider moving troops from the Balkans to western Europe when an event occurred quite outside the scope of anything the deception staffs could devise. The opening of the Soviet offensive on the southern part of the Eastern Front on 4 March led instead to the transfer of formations to stem this new red tide. Meanwhile, the effect of WANTAGE, the latest in the series of false order-of-battle plans, began to pay dividends. By May OKW was seriously alarmed. Jodl's deputy, the newly promoted General der Artillerie Walter Warlimont, produced an appreciation claiming that 'unquestionable' sources confirmed the Allies planned to assault the Balkans from Egypt and Libya.[32] In the Mediterranean theatre the overestimation was as much as eighty-five per cent and played a major part in preventing the movement of German troops from that theatre to the crucial arena of northern France.[33]

Unlike in Scotland, however, matters in the Mediterranean were complicated by plans for genuine operations. At Tehran it had been agreed that an invasion of the south of France would be mounted simultaneously with OVERLORD. Appropriately (but rather insecurely) code-named ANVIL, this operation was renamed DRAGOON and postponed owing to a shortage of landing craft and the lack of success in Italy, including the landings at Anzio. A threat to German forces in the south was desirable to assist operations in the north, but obviously this would complicate matters when DRAGOON with its associated deception plan, FERDINAND, went forward. 'A' Force was given the task of keeping these German reserves away from Normandy until D+25 under the code-name VENDETTA. A threat was developed from US Seventh Army in North Africa to the Narbonne region, conveniently well clear of the Toulon-Nice area where the genuine landings were due to take place. German sources reveal only partial success for VENDETTA, although ironically FHW drew attention to the notional visit of Montgomery, suggesting that it might presage operations in the south. In fact, this was Lieutenant Clifton James, taking part in Operation COPPERHEAD, intended to draw attention away from impending operations in northern France.

Although during the first weeks of the Battle of Normandy an invasion of the south of France was not seen by the Germans as imminent, preparations were nevertheless made. (Although it reported increasing activity in North African ports, OKW saw any such operation as being subordinate to OVERLORD). Heeresgruppe G deployed ten divisions, including two panzer divisions, on 6 June, and not until mid-June was a panzer division moved north; and no others were moved until mid-July. German forces were therefore successfully diverted from the critical theatre.[34] With the winding up of VENDETTA, Wilson, now Supreme Allied Commander Mediterranean Theatre, was able to report on 26 June that:

> No [enemy] divisions moved from the Mediterranean theatre to north-west Europe during the preparation period of OVERLORD. ... So far as is known to date, only one division has moved from the Mediterranean theatre towards the Overlord area; and none arrived in time to influence the battle during the 'critical' period defined by SHAEF.[35]

One other deception plan was made to keep reserves away from northern France, although it was conspicuously unsuccessful. Operation IRONSIDE was designed to threaten the western French coast. The cover story was that a largely American expeditionary force was concentrating in western UK ports and aimed to capture Bordeaux, which would then receive forces direct from the USA. The deception staffs had little confidence in this tale and with few physical resources to devote to it the only means to disseminate it were the double agents, whose controllers were reluctant to risk precious credibility. Some reports were sent but there is no evidence of their being accepted.[36]

OPERATIONAL DECEPTION: FORTITUDE SOUTH

That Hitler's attention was focused on north-west France had been obvious since his issue of Directive No. 51 on 3

November 1943, describing that theatre as his top prior-ity for the coming year.[37] While demonstrating the prob-able futility of attempts at grand strategic deception, it highlighted the importance of the Pas de Calais in Hitler's thoughts, particularly in relation to the V-weapons, whose emplacement was then beginning and whose limited range demanded close proximity to Great Britain. This would also prove of vital significance to the deceivers when the battle came to be fought. Not only did the Pas de Calais present the Allies with the shortest route across the Channel, but it would also provide the quickest way into the industrial heartland of the Ruhr and of Germany itself, a fact well known to the Germans. It was also logical for the Germans to assume that the British and Americans would want to neutralize the V-weapon sites before they could do serious damage to London and the Channel ports, especially given the propaganda about their value that Goebbels was producing.

In January 1944 Rommel, who had been dispatched to inspect the western defences, proposed to Hitler that radi-cal changes be made to von Rundstedt's anti-invasion plans and was granted his request of command of Heeresgruppe B, covering the critical sector from the Zuider Zee to the Loire. Nominally subordinate to von Rundstedt, Rommel held radically different views from his superior on the problem of defending 1, 700 miles of coastline. This reordering of the High Command in the west produced further complications. All armoured forces now came under the control of General der Panzertruppen Baron Geyr von Schweppenburg, in command of Panzergruppe

West. He wished to keep these forces concentrated and was backed up in this belief by the Inspector-General of Panzer Troops, Generaloberst Heinz Guderian. Rommel's experience of growing Allied air power convinced him that if they succeeded in effecting a lodgement, the Allies would be simply too strong to evict. He wanted to establish strong tactical reserves close to the beaches and defeat the invasion there. Most accounts portray the division as being between von Rundstedt and Rommel, but it was principally between Rommel and von Schweppenburg supported by Guderian, with von Rundstedt as arbiter. Eventually, Hitler produced a compromise that satisfied nobody, ordering that four panzer divisions be kept as an OKW reserve.[38]

Meanwhile, von Rundstedt, who in common with most of the German Army was unable fully to comprehend the air and naval dimension of the invasion, assumed that similar air and naval weaknesses to those that had compelled his own cross-Channel plan of 1940 to take the shortest route would apply equally to the Allies, a belief that further highlighted the Pas de Calais. As a result of the ill-fated Dieppe raid of 19 August 1942, German commanders at all levels had drawn the wrong conclusions. The Dieppe operation confirmed their belief that an invasion would target a major port. Rommel at first believed the Allies would land at the mouth of the Somme and try and take Le Havre from the rear.[39] However, despite being repulsed with heavy losses, the largely Canadian force had given the Allies 'the priceless secret of victory'.[40] One of the most important lessons learned was the impossibility of quickly capturing a port intact. The ingenious solution was provided by the Mulberry harbours, consisting of enormous floating concrete caissons to be joined together on the far side of the Channel,* and Pipeline Under the Ocean (PLUTO) for supplying the vast quantities of fuel that would be needed for the fully motorized forces. At the same time, by negating the need to capture a major port immediately, these innovations would also greatly assist deception planning.[41]

Thus von Rundstedt, like Hitler, concentrated his focus on the Pas de Calais. Consequently, the labourers of the Todt Organization built 132 heavy concrete gun emplacements between Dunkirk and Boulogne, compared with 47 in Normandy. Goebbels' ministry claimed: 'We have fortified the coast of Europe from the North Cape to the Mediterranean and installed the deadliest weapons that the twentieth century can produce.' Gilles Perrault, historian of the French Resistance, noted that 'if the Atlantic wall anywhere resembled the ferocious image popularized by Goebbels, it was in the segment where von Rundstedt expected the Allies.' Rommel agreed, and told Guderian repeatedly that the invasion would occur north of the Somme.[42] With characteristic vigour, he had over four million mines and thousands of obstacles, many booby-trapped (known as 'Rommel's asparagus') placed along the foreshore. Being a camouflage enthusiast

---

* Each hollow concrete caisson was 230 feet long by 60 feet across, and therefore impossible to conceal. The story was put about that they were boom defence units for the protection of harbours (PRO WO 219/2237).

himself, he also added numerous dummies and decoys which drew considerable attention from Allied airmen.

Commanding the land component of the invasion force was Montgomery. His intelligence sources estimated a total of sixty German divisions in France, including ten panzer and twelve mobile infantry divisions.* The Allies had a total of thirty-seven divisions available in the landing and follow-on forces. Apart from the immediate threat posed by the five divisions *in situ* at the landing sites, there were four panzer divisions close enough to affect operations in the area. Montgomery's projected buildup measured against that of the Germans in attempting to contain and repel the Allies led to the conclusion that after forty-eight hours there would be approximately twelve divisions on either side, by which time the Germans would realize the seriousness of the threat.[43]

The actual assault, while in itself extremely hazardous, would not thus ultimately decide the issue; instead, the race to build up forces in the lodgement area would be the critical factor, and in this the Germans had the advantages of numbers and land communications. Deception would play a crucial role in producing a ratio of forces necessary for Allied victory in the battle of the build-up

---

* There were in fact fifty-eight. 'Most of them were low-grade divisions, and some were skeletons,' according to von Rundstedt (quoted in Liddell Hart, *The Other Side of the Hill,* p.382). In total: thirty-three were static, thirteen mobile infantry divisions (i.e., having the full complement of divisional transport, much of which was in any case horse-drawn), two *Fallschirmjäger* (parachute), and six panzer divisions (plus the OKW reserve of four panzer/panzergrenadier divisions; see H. Meyer, p.32).

and permitting a break-out. The cover plan of FORTITUDE SOUTH was simple. The Allied main assault would be launched against the Pas de Calais six weeks after the landings in Normandy which were themselves no more than a diversion to draw off reserves. The aim was thus nothing short of pinning the nineteen divisions of Fifteenth Army in the Pas de Calais. Its success in misleading every level of the German Command as to the precise target and strength of the threat was one of the central triumphs of the deception staffs of the entire war.[44]

The use of air power was critical to the deception plan. Control of the skies not only provided security for the preparations but formed an integral part of them. Now the air plan was framed so that for every ton of bombs dropped on coastal batteries and every reconnaissance mission flown over Normandy, two would occur in the Pas de Calais. The region north and east of the Seine received ninety-five per cent of the anti-railway bombing effort.[45] This was a vital aspect of the operational plan, since Eisenhower insisted that he could not outstrip the German build-up without paralysing rail communications.[46] Fortunately, it also coincided with the deception plan. Eisenhower's team of railway experts produced a target list that attacked repair shops and junctions between the Seine and the Meuse, but the lines serving Normandy were mainly extensions of those supplying the Pas de Calais, which therefore still appeared the target. The campaign opened in March and was hugely successful in both objects. Barely forty-eight hours before D-Day, in his weekly report to von Rundstedt, Rommel stated that 'concentrated air attacks on the coastal defences between Dunkirk and Dieppe [i.e., the Pas de Calais] strengthen prospects of a large-scale landing in that area'.[47]

The main deceptive elements were less easily formalized. The proposal of a diversionary operation was ruled out on the grounds that only a genuine landing would bring the Luftwaffe to battle. There were only enough landing craft available to land one division in the Pas de Calais and so puny a landing would quite obviously be a diversion, with disastrous consequences for the genuine assault. Given that this would land between the Seine and the Cotentin peninsula, these forces would be assembled in the area stretching from Portsmouth westwards as far as Plymouth. The preparations could not hope to be concealed permanently and were already apparent to FHW. Instead, it was planned to produce equal concentrations in south-east England, which would be intensified once the actual landings had taken place. When the demands this would place on scarce resources became apparent, it was further decided, given almost total Allied control of the air and of the tried and trusted system of double agents, to rely mainly on the latter and on radio simulation, since by now these were what the Germans almost exclusively relied upon. Nevertheless, visual simulation would be needed as back-up, notably of landing craft in the Kent ports and Thames estuary, because there could be no guarantee that aerial reconnaissance would not get through and the plan could not afford a gap in the corroborative evidence that might imperil the entire operation.[48]

The knowledge the Allies possessed of German intelligence capabilities and perceptions now proved invaluable. Von Roenne at FHW was known to be a shrewd and experienced staff officer from a Prussian *Junker* family, whose understanding of the Allies' true capabilities and intentions was often remarkably accurate. He was one of the first to see the Norway threat as a deception and never took the threat to the Balkans seriously. But he was dismayed to find the High Command put far more faith in Hitler's intuition than in genuine intelligence. Well aware that a major invasion was imminent, he was alarmed at the way the West was used as little more than a rest and recuperation theatre for shattered formations pulled out of the line in the East, and was determined to make the threat clear for his superiors.

Reports from the German radio-direction finding service, the Y-Dienst, showed that while traffic in the Mediterranean was decreasing, in the United Kingdom it was steadily increasing. Certain indicator formations were carefully monitored, including the US 82nd Airborne Division. Although it appeared to be still in Italy, careful analysis of the rhythms and patterns of its operators showed changes that suggested a deception, and this seemed to be confirmed when new patterns in England fitted the 'signature' of the division. Confirmation came in a signal intercepted in clear; a routine message concerning a welfare matter but referring to the 82nd command post at Banbury.

Now the rivalry between the various German intelligence agencies handed the LCS a new ace. Von Roenne discovered that Kaltenbrunner's SD was scaling down his estimates of Allied strength, often by as much as fifty per cent, in order to tell Hitler what he wanted to hear and cover its own lack of information and inefficiency. One of von Roenne's staff suggested deliberately overestimating in order to strike a 'balance'. Von Roenne was either initially horrified by this suggestion or, as an intimate of various members of the July bomb plot, was keen for an Allied landing to counter-balance the Soviet threat. Either way, he realized he need only accept information at face value without seeking corroboration, and a consistent picture would emerge. This naturally played directly into the hands of the deceivers, but hardly had von Roenne begun issuing his overestimates when for reasons of his own Kaltenbrunner stopped scaling his down! From May 1944 the fictitious order of battle comprising some ninety divisions (where in reality there were just thirty-seven) became set in stone.[49] Even when von Roenne began to have misgivings about this situation, it was difficult to see what to do about it.

Meanwhile, Montgomery's Chief-of-Staff, Francis de Guingand, had raised the importance of the post-assault phase, saying: 'I feel we must, from D-Day onwards, endeavour to persuade [the enemy] that our main attack is going to develop later in the Pas de Calais area, and it is to be hoped that NEPTUNE will draw away reserves from that area.'[50] As the Germans would be bound to wonder why the Allies held back such powerful forces, they could be further diverted after the initial landings from the crucial area of operations. This would make use of the entirely notional First US Army Group (FUSAG), which had come into being in October 1943 and would now come under the command of the larger-than-life Patton. The latter was under a cloud after a number of scandals involving the press, but the situation provided the Allied High Command with a means to utilize Patton's public talents before he was due to take command of the US Third Army later in the campaign. Characterized by Ladislas Farago as 'the tall, taut, tense American general of vague fame and growing notoriety, the swashbuckling tank wizard the tabloids in the States already called "Old Blood and Guts"', Patton was the senior Allied commander most respected by the German generals (if not by Hitler, who contemptuously referred to him as the 'cowboy general'). In their eyes he was a natural choice to command the main effort of the Allies and therefore an ideal choice for the deceivers. This aspect of the plan was code- named QUICKSILVER and divided into six parts.*

Although deeply disappointed not to be taking part in the invasion plan itself, Patton realized the importance of his role and embraced it with gusto under the direction of Lieutenant-Colonels John Jervis-Reid and Roger Hesketh. German reports began to refer to Armeegruppe Patton,

---

* QUICKSILVER I was the fiction that the assault on the Pas de Calais would be launched some weeks after the Normandy landing. QUICKSILVER II covered the radio deception, QUICKSILVER III the display of landing craft, QUICKSILVER IV and V the bombing campaign, and QUICKSILVER VI the display of sixty-five false lighting schemes along the south coast, including an exact replica of the port and railway lighting of Newhaven at Cuckmere Haven.

and his later absence from Normandy helped reinforce the idea that the landings were diversionary. Genuine formations were included in FUSAG, to be replaced by notional ones upon their transfer to France, including many elements of Fourth Army from Scotland. The principal components were US Third Army, Canadian First Army and dozens of US divisions still in training in America. According to the cover plan, twelve divisions were to take part in the 'Calais landings' with thirty-eight waiting to follow up, all covered by US Ninth Army Air Force. The careful blend of real and imaginary formations thus served a threefold purpose: cloaking the build-up, disguising the destination and strengthening the credibility of the cover plan.[51] When Patton finally went overseas, he was replaced by Lieutenant-General Leslie McNair, and the deception was eventually maintained far longer than originally intended.

As with Fourth Army, a carefully prepared programme of radio traffic commenced on 24 April to simulate the busy network of an army group in preparation for major operations. A special Royal Signals unit, 5th Wireless Group, introduced a number of original features. All traffic was recorded in advance and equipment was devised that made it possible for one transmitter to simulate six, so that the complete traffic of a divisional headquarters and its brigades could be broadcast by a single radio truck. Unlike 5th Wireless Group, the US 3103rd Signals Service Battalion used normal equipment, and the traffic passed directly between operators without the intervention of recording apparatus, with the trucks moving through the country just as the formations which they represented would have done. By omitting links below the divisional level they were able to represent no fewer than three army corps and nine divisions as well as the headquarters of an army group and an army.[52] Having arrived in February, the 3103rd spent their time carefully analysing formation and unit signals patterns in order to reproduce them accurately. Security was of paramount importance and a directive to signals personnel stated:

> You are taking part in an operation . . . designed to deceive the enemy, and it has a direct bearing on the success of our operations as a whole. You must realize that the enemy is probably listening to every message you pass on the air and is well aware that there is a possibility that he is being bluffed. It is therefore vitally important that your security is perfect; one careless mistake may disclose the whole plan.[53]

One such message became a SHAEF classic: '5th [Bn] Queen's Royal Regiment report a number of civilian women, presumably unauthorized, in the baggage train. What are we going to do with them – take them to Calais?'[54]

The radio deception was once again backed up with appropriate visual displays, although the almost total command of the skies meant that wholesale layout of dummy camps and depots was deemed both unnecessary and impractical. Twenty-First Army Group never agreed

to the extensive programme of physical deception originally proposed by SHAEF. The enemy would monitor the
shipping in ports rather than the camps around them as an
indication of where the attacking forces would come from
and of the scale of the attack. The enemy had made no
inland aerial recce and even if he did, owing to the character of the British landscape with its woods, towns and
villages, he would be unlikely to discover much if camouflage discipline was strictly enforced, as he would expect
it to be. These costly measures would do little good and
might do much harm, and 21st Army Group requested
that SHAEF cancel any instructions that the latter had
issued in connection with camouflage, meaning the abo-
lition of 'discreet display' in Eastern and South-Eastern
Commands and the practice of normal camouflage discipline in all areas.[55]

Displays were instead restricted to shipping. Fake vessels resembling Landing Craft Tanks (LCTs) were known
as 'big-bobs', other inflatable dummy craft as 'wet-bobs'
(a reference in the 'old boy' world to those at Eton who
rowed for their house or school). Some were moored in the
creeks and inlets between Great Yarmouth and the Thames
estuary, but the larger proportion of them were concentrated around Kent and eastern Sussex.* Attention to detail
ensured authenticity from the air, even down to oil spills

---

* By 12 June 255 dummy craft had been put on display by men of 4th Bn,
Northamptonshire Regiment, and 10th Bn, Worcestershire Regiment,
and 'operated' by naval personnel at Great Yarmouth, Lowestoft, in the
rivers Deben and Orwell, and at Dover and Folkestone (M. Howard,
p.127; G. Hartcup, pp.89–90).

and suitable lighting at night to suggest loading operations. Most spectacularly, the designers of Shepperton Film Studios were enlisted to build a giant 'oil storage facility' and docking area near Dover, designed by Basil Spence. This stretched several miles and was complete with storage tanks, pipelines, jetties, terminal control points and anti-aircraft defences. Both the King and Montgomery paid visits, which were prominently covered in the press. An RAF photo-recce plane overflew the complex, and modifications were made from the resulting prints. Thereafter the defensive screen was gradually relaxed, although strong anti-aircraft defences kept those German aircraft that did come across at high altitude, from which it was difficult to pick out any remaining flaws. When the Germans fired a few long-range shells from Cap Gris-Nez, the camouflage crews even produced suitably realistic damage using sodium flares and smoke generators.[56]

Security was as vital to visual deceptions as to radio transmissions. Jim Rowe was working on the railway between Reading and Guildford when he realized that the load his engine was pulling was extremely strange. He said to his mate: 'Here Bill, these tanks are made of wood.' However, to German reconnaissance planes, 'they *were* tanks – there wasn't no argument about it. They were wonderful fakes, right down to the guns on them and the dark green camouflage.' Well aware of the wartime doctrine that 'Careless Talk Costs Lives', Jim and Bill kept quiet about their unusual load.[57]

Security proved something of a headache for the planners, however, since there were a considerable number of interests to accommodate, many of them conflicting. The Foreign Office wanted to retain normal diplomatic channels while the deception staffs wished to ensure, and if possible increase, the plausibility of their schemes. The proposal that all communications from Britain be cut at the last possible moment was rejected by Bevan on the basis that such an action would alert the Germans at the most critical moment. He would rather risk leaks, which would in any case be low-level and contradictory and which he felt confident the LCS could smother, if such a communications ban could be imposed early enough to cover both FORTITUDE SOUTH and NEPTUNE. This would mean letting certain civilian authorities in on some aspects of the plan to enable their co-operation so that civil measures of security, civil defence and evacuation could be instituted. However, once the decision had been taken to rely on radio deception and 'special means' (the double agents), these measures were deemed largely unnecessary.

At the beginning of February Churchill insisted that security measures should go 'high, wide and handsome'. All service leave, travel to Ireland and the airmail service to Lisbon were suspended and the coasts declared prohibited to a depth of ten miles. The suspension of the Lisbon mail was a serious inconvenience to the Twenty Committee, since their principal double agent, GARBO, could no longer send long, rambling letters. Once imposed, it was important that these measures should remain for the longest possible duration, which was vehemently opposed by the Foreign Office. The Chiefs of Staff agreed to relax it on D+2 (in practice, D+7) but this in turn presented the deception planners with the thorny conundrum of explaining why severe security was imposed for the landings in Normandy but relaxed before the supposedly imminent assault on the Pas de Calais. A clever scam was devised whereby GARBO, 'informed' by his contact in the Ministry of Information, 'learned' that although the military had wished to continue with the ban, it would have meant telling neutral diplomats about the Pas de Calais landings and that this was deemed a greater security risk. However, this too was rendered unnecessary when Eisenhower insisted that the diplomatic ban remain in force until D+25, and security for FORTITUDE SOUTH remained intact to the very end.[58]

Many other seemingly trivial elements contributed their part to the overall picture. Naval craft were busy off East Anglia and in the Channel and electronic warfare units jammed Nazi radar, but for all the efforts of the deceivers working to create the visual and electronic image of FUSAG the most important source of information as far as both sides were concerned were the double agents. Of these, the most significant was undoubtedly GARBO. The young Spaniard had initially been rejected when he approached the British, so he enlisted in German service under the code-name CATO before offering himself to the British once more as a ready-made double agent. He in turn operated a network of fourteen agents (including 'Welsh nationalists', many totally fictitious) in places such as Brighton, Dover and Harwich. Between January 1944 and D-Day he averaged over four transmissions a day. Beginning with an unsuccessful attempt to sow seeds of doubt as to the arrival at all of the Allies, GARBO along with BRUTUS, TRICYCLE, TREASURE and TATE maintained a steady stream of information that regularly found its way into German intelligence summaries. As the big day approached, so the arrangements for the double agents became increasingly tightly orchestrated. Information about the bogus structure of FUSAG was mixed with information about Montgomery's genuine 21st Army Group, which when these formations were recognized in Normandy would strengthen the credibility of the fake reports.

At the beginning of May GARBO reported the presence of US 6th Armored Division in Ipswich. Unbeknown to the deceivers, German aerial recce had observed the strong build-up in the south-west, and von Roenne was worrying about this apparent shift. GARBO's message, together with a number of others, helped restore German faith in their original estimate. On 12 May von Roenne issued a report suggesting the bulk of forces were in the south-west and only diversionary operations were expected from the south-east, exactly the opposite of what the deceivers wanted. After further messages from various agents and increased air activity off the south-east coast (actually intended to prevent the Luftwaffe from checking on the presence of the fictitious units) von Roenne changed his mind. A map, subsequently captured in Italy and dated 15 May 1944, showed the completeness with which the Germans accepted the imaginary order of battle, particularly as a result of the reports of GARBO and BRUTUS.[59]

While it proved enormously successful, the double-agent system was fraught with terrible risks.

One opportunity arose, however, to show the build-up in the 'south-east' to someone entirely trustworthy to the Germans. Throughout the war prisoners were repatriated under the auspices of the Swedish Red Cross in cases of serious illness. General der Panzertruppen Hans Cramer, holder of the Knight's Cross of the Iron Cross with oak leaves and one-time commander of the Deutches Afrikakorps, was being held in a camp in south Wales when he became a candidate for repatriation. Immediately the LCS took control, arranging for his transport by road to London. It was 'let slip' that his route was through southern and south-eastern England when in fact it went through the areas of maximum concentration in the south-west. The removal of all signposts years earlier assisted the illusion. After dinner with Patton, who was introduced as Commander-in-Chief of FUSAG, and who discreetly mentioned Calais, Cramer was dispatched to Germany by means of the Swedish ship *Gripsholm* and arrived on 23 May in Berlin, where he duly reported that south-east England was awash with troops and equipment bound for the Pas de Calais.[60] His reports were received with some dismay in Berlin, where Goring accused him of adopting a defeatist attitude, but after a few days leave he was posted to von Schweppenburg's staff in Paris.[61] The seeds of misinformation had been planted and watered and were finally beginning to germinate, but the green shoots of deception were incredibly fragile, and needed all the protection that security and luck could give them.

A series of great scares now alarmed the Staffs, beginning with the discovery of a parcel in Chicago's Central

Post Office containing precise and detailed instructions for the invasion. Then it was feared that, during the German E-boat raid on Exercise TIGER in Lyme Bay, one of a number of missing officers, fully briefed on the invasion plan, might have been captured. A thorough and detailed search of the waters around the bay was conducted until each body had been accounted for. Furthermore, reports from the French Resistance told of the arrival from Hungary of the crack Panzer Lehr Division and its deployment at Chartres and Le Mans, barely 100 miles from the beaches, which 12th SS Panzer Division Hitlerjügend also menaced from Lisieux. The move of 21st Panzer Division from Rennes to Caen, and the addition of 6th Parachute Regiment and 91st Light Division to the critical sector at the base of the Cotentin peninsula were especially alarming: it appeared that the Germans were somehow aware of Allied plans.

Throughout the spring arguments had raged within the German High Command as to precisely where and what form the landings would take. Von Rundstedt was not alone in being convinced that any invasion would come across the Straits of Dover; in this conviction Hitler for the most part agreed, but the increased strength in Normandy came as a result not of intelligence on the Germans' part but of intuition by Hitler. His mind did not operate along the lines of the trained staff officers around him, and a naval report on 26 April suggested that the pattern of Allied air attacks and minesweeping operations indicated an attack between Boulogne and Cherbourg. Whatever his influences, on 2 May Hitler proclaimed to his staff that the Allied *schwerpunkt* would be in Normandy. He reasoned that diversionary attacks were likely to try and draw off von Rundstedt's reserves, and that the Allies would need a big port situated in such a way as to be defensible from a short line – conditions met by Cherbourg and the Cotentin peninsula. Fortunately for the Allies, according to Warlimont, 'he believed furthermore, previous to and for a long time after the invasion, that a second landing would take place on the Channel coast.'[62] Thus the basic tenet underpinning all the deceivers hopes in fact held firm in the minds of the German High Command. Hitler had issued orders for strengthening the lines between the Seine and the Loire, but by insisting that all of France be held, he restricted his subordinates' freedom of action. Von Rundstedt, his gaze fixed firmly on the Pas de Calais, carried out these instructions reluctantly.

If the Germans thought they knew the place of the invasion, they also tried to second-guess the Allied commanders as to its timing. Believing that the assault troops would not wish to cross hundreds of yards of exposed beach, they assumed high tide as a logical prerequisite, combined with dawn and reasonable weather. An alert was held on 18 May, during which all leave was cancelled and tension rose. The Kriegsmarine experts consulted their tide tables, then proclaimed that no real danger now existed until mid-August. Leave, training and exercises such as the one planned for senior officers in early June could all be resumed. The Allies were indeed ready and the real date had by now been set, but they would land at dawn on a spring low tide in order to find Rommel's obstacles exposed rather than have them rip open the hulls of landing craft. A rising moon and long summer nights would give maximum assistance in the build-up to support the initial waves of parachutists and help to provide naval gunfire support.[63] Only the notoriously fickle early June weather could interrupt proceedings and in this a risk had to be taken.

Originally planned for May, the operation was finally scheduled for 5 June but was postponed for a further twenty-four hours owing to a severe storm. When the meteorologists indicated that despite its apparent worsening there would in fact be a narrow gap sufficient to launch NEPTUNE, Eisenhower spoke the famous phrase, 'Okay, let's go'. Without the benefit of weather data from the far west, the German meteorological service had failed to spot the break and had declared conditions would be too bad for landing operations throughout the first week of June. As a result, many senior officers, including Rommel, took the opportunity to take leave. Warlimont noted in his diary that nobody in OKW had the slightest suspicion that 'the decisive moment of the war was upon them'.[64] Early indications that something was afoot were ignored by von Rundstedt's headquarters, but Rommel's sent a 'Most Urgent' signal to stand-to at 2300 hours on 5 June. However, this was sent to Fifteenth Army between the Scheldt and the Orne; Seventh Army, guarding the actual invasion beaches, received no warning at all.

Once the vast resources assigned to it were unleashed it would be impossible to conceal the invasion for long, but the aim of the deceivers was to ensure that the Germans were equipped with powerful radar-jamming equipment. However, jamming a sector of coast would be a clear indicator of the direction of attack and it was impossible to jam the entire coastline, which would in any case put all defences on alert. It was necessary to give the Germans some information, but it had to be false information. A force of heavy bombers fitted with powerful MANDREL jammers was sent to block the early warning radars east of the Seine beyond the Belgian border. Two gaps would be left, apparently by error or atmospheric conditions, and deception fed through them. The western gap was near Le Havre, to the east of the landing zone. Here a Kriegsmarine long-range SEETAKT set was capable of picking up shipping in the Channel. No. 617 Squadron RAF – the famed 'Dambusters' – was tasked to fly Operation TAXABLE, a precision pattern more complicated than anything previously attempted. Taking account of the exact characteristics of the SEETAKT, they would paint on its screens a picture of hundreds of ships, which were simply unavailable in reality. To do this, WINDOW would have to be dropped in exactly the right density and shape, a process that involved turning repeatedly to port, flying a set distance and turning again, with one flight eight miles behind the other, advancing the screen so that it appeared to be moving towards the coast at a steady eight knots. The crews had to be relieved every two hours, which introduced the additional difficulty of taking over in just the right position, and there was a serious risk of collision in the moonless sky.

The ruse worked and brought the expected response: radar-equipped aircraft were sent to investigate. The second phase of TAXABLE was designed to account for this. Some naval launches were detached from the fleet and continued up the Channel past Le Havre, sailing within the frame created by the WINDOW and towing behind them balloons called FILBERTS fitted with radar reflectors. Others carried MOONSHINE, and audio effects were added so that the picture was created of a large fleet actually moving away from Normandy. The Germans turned naval craft, patrol aircraft and shore batteries on TAXABLE but none on NEPTUNE. At the same time No. 218 Squadron was flying a similar mission close to the Pas de Calais in Operation GLIMMER, which also drew some limited response.

Meanwhile, the extremely vulnerable transport planes carrying the first wave of the assault, the airborne divisions, were approaching the coast and it was imperative that German night fighters be kept well away or disaster would ensue. Some twenty-four Lancasters and five Flying Fortresses from 100 Group RAF flew a mission that simulated a vast bombing raid along the Somme, dropping vast quantities of WINDOW as they went and at the same time interfering with ground-control and fighter communications.[65] The German radar screen was therefore not only neutralized, but acted as a conduit for deception and helped preserve the element of surprise, so that the Germans received no warning of the air armada bringing the three airborne divisions. The Luftwaffe, practically deaf and blind, instead spent the crucial hours between 0100 and 0400 hours searching for a non-existent stream of bombers apparently operating over Amiens.

During this time reports were coming in from all over Normandy and Fifteenth Army's area of parachute landings. In Operation TITANIC a series of drops was made by small groups of the Special Air Service accompanied by dummy parachutists called PARAGONS.[66] The first drop came a few minutes after midnight near Marigny, between Coutances and St Lô, and the others were made in a broad sweep right around the landing area from almost the west coast of Cotentin to just east of Le Havre. The dummies were designed to fire battle simulators on landing, to reproduce the sound of small-arms fire, mortars and grenades; pintail bombs fired parachute flares and Verey lights and the four four-man parties operated gramophones with records of soldiers' voices and more battle sounds. Even canisters of chemicals to reproduce the smells of combat were used. The aim of these drops was to delay enemy reserves from reaching the real drop zones (DZs) and to cause maximum confusion in the rear areas. The scattered nature of the genuine drops, in which many men missed their designated DZs (especially Americans around the base of Cotentin), inadvertently added to the effect and began to dissipate local reserves, which were dispatched hither and yon. Soon von Rundstedt's operations map was a rash of red spreading far beyond the actual area of battle, making the *schwerpunkt* impossible to identify and the flanks a blur. In this respect TITANIC was very successful, and one important German regimental group found itself fruitlessly sweeping woods at Isigny through the early hours of the morning, well away from the real landings to their north.[67]

Defending OMAHA beach was the 352nd Infantry Division, commanded by Generalleutnant Helmut Kraiss. The half-mile of broad flat beach at low tide was edged by a bank of shingle three to four feet high, behind which lay a marshy strip and then a steep ridge between sixty and eighty feet high. From these heights, which were not effectively neutralized by the initial bombardment, Kraiss's two forward regiments gave the beach its nickname 'bloody Omaha'. However, apart from pinning the Americans on the beach, Kraiss was in no position to throw them off it because he had no effective counter-attack force. His division provided the corps reserve, Kampfgruppe Meyer, comprising three battalions. From 0305 hours it deployed westward towards Isigny and Carentan to open the roads and search for parachutists, east of the US 101st Airborne Division's actual DZ. Once committed, it was practically impossible to recall and, although one battalion was directed to counter-attack towards Colleville-sur-Mer at 0735 hours, radio communications were proving difficult and orderlies had to be dispatched by bicycle to round up the detachments. Having little motor transport, nothing had come of this order four hours later. The rest of the Kampfgruppe was in no better state to intervene when that decision was finally made, and by late afternoon it had been broken up and effectively destroyed from the air. By this time the crisis on OMAHA was resolved and the Americans were able to push more than a mile inland. TITANIC thus not only distracted the Germans from the airborne DZs but played a vital part in enabling success at OMAHA, without which an extremely dangerous gap could have been opened between the British and American zones.

Normandy and Beyond

Following the leading British troops ashore came Colonel David Strangeways' 'R' Force to provide tactical deception measures in subsequent operations. These began with Operation ACCUMULATOR, on 12 June. Tactical deception had played an enormous part in the success of the landings themselves but the crucial factor in deciding the successful outcome of OVERLORD would be the battle of the build-up, which now reached its climax. FORTITUDE SOUTH continued, principally through the double agents. GARBO was able to improve his credibility by sending a message giving the details of NEPTUNE just too late to be of any use. No one was listening at the time, so later he berated his control for being unavailable at the crucial moment. He now proceeded to send prolific reports and spent a number of notionally sleepless nights. Montgomery's race to win the battle of the build-up was now in full swing. As early as 0400 hours on the morning of the invasion von Rundstedt's Chief-of-Staff, Gunther Blumentritt, asked OKW for the release of the armoured reserves. Jodl, speaking for Hitler, refused, saying significantly that the landings were no more than a feint and that another landing would come east of the Seine. This argument continued all day. Hitler had gone to bed with a sleeping draught and amazingly he was not informed of the seaborne landings until mid-morning. His hopes of throwing the invasion straight back into the sea evaporated and he sent orders to Fifteenth Army to send reinforcements to Normandy.

No fewer than five infantry divisions were set to move to the Seine bridges, and two panzer divisions, including 1st SS Panzer Division Leibstandarte Adolf Hitler in Belgium, were to move to the bridgehead. On the night of 8 June GARBO sent the following message:

> After personal consultation on 8 June in London with my agents, JONNY [*sic*], DICK and DERRICK whose reports I sent today I am of the opinion, in view of the strong troop concentrations in SE and E England, that these operations are a *diversionary manœuvre designed to draw off enemy reserves in order to make an attack at another place*. In view of the continued air attacks on the concentration area mentioned, which is strategically favourable for this, it may very probably take place in the Pas de Calais area, particularly since in such an attack the proximity of air bases will facilitate the operation by providing continued strong air support.

This was passed at 2200 hours on 9 June to Oberst Friedrich-Adolf Krumacher, the OKW Chief of Intelligence, who added that it 'confirms the view already held by us that a further attack is expected in another place (Belgium?)' and the message was seen by Hitler.[68]

Once the Germans had identified elements of 21st Army Group in Normandy, confirming GARBO's earlier information, his credibility was such that his information on FUSAG was accepted completely. A stroke of luck, in the form of a message from an uncontrolled agent in Stockholm, JOSEPHINE, referring to rumours in London of an attack on the Pas de Calais, was enough to generate the following message from von Rundstedt on 0730 hours the next morning: 'As a consequence of certain information, OB West has declared a "state of alarm II" for the Fifteenth Army in Belgium and North France. (For Netherlands command too, if Heeresgruppe B thinks fit). The move of 1st SS Panzer Division will therefore be halted.' The Allies were able to watch German reinforcements move towards the battle area, halt, then turn around and move away again. FORTITUDE SOUTH was maintained for far longer than originally intended, and only slowly did the Germans release formations from Fifteenth Army into the battle raging to the west. Fourth Army was notionally brought south to support FORTITUDE SOUTH as FORTITUDE NORTH was wound up. On 18 July von Rundstedt's headquarters sent a message, intercepted and read at Bletchley, stating that 'there are no grounds for changing our appreciation of the intentions of the Army Group assembled in south-east England'. Only by the end of the month were the Germans seriously dubious about the likelihood of a further assault on the Pas de Calais, and this was overwhelmingly because they believed the Allies simply wanted to reinforce the success they had achieved in Normandy and not through any suspicion that FUSAG did not exist. On 3 August its commander, Patton, opened his genuine offensive to the landing in Normandy, however, the Germans would be the victims of another massive deception operation that would lead to their biggest single defeat of the war, far more destructive than that of Stalingrad: Operation BAGRATION.

CHAPTER TEN

MASKIROVKA

'... the Soviet armed forces learned to preserve in deep secretiveness the intentions to execute disinformation on a large scale and to deceive the enemy.'

Marshal G. K. Zkukov

OPERATION BAGRATION WAS launched just over two weeks after D- Day, on the third anniversary of the German invasion of the Soviet Union, 22 June 1944. By the close some two months later Germany had suffered its most crushing defeat of the war – the complete destruction of Army Group Centre, involving the loss of some 350,000 men – and the Soviets had advanced menacingly close to the borders of the Reich. A major factor in Soviet success was their use of *maskirovka* to achieve surprise. The *Soviet Military Encyclopaedia* defines *maskirovka* thus:

The means of securing combat operations and the daily activities of forces; a complexity of measures, directed to mislead the enemy regarding the presence and disposition of forces, various military objectives, their condition, combat readiness and operations, and also the plans of the commander ... *maskirovka* contributes to the achievement of surprise for the actions of forces, the preservation of combat readiness and the increased survivability of objectives.[1]

It permeates down to the lowest tactical level and includes all measures, active and passive, designed to deceive the enemy. Although the word is sometimes translated as 'camouflage', this belies its much broader meaning which includes: concealment (*skrytie*), imitation using decoys and dummies (*imitasiia*), manœuvres intended to deceive (*demonstratinvnye manevry*) and disinformation (*dezinformatsiia*).[2]

Although a product of the Soviet era, *maskirovka* has a long history; its roots can be traced to the Imperial Russian Army, and several Soviet authors traced it to Dmitry Donskoy's placing a portion of his forces in an adjacent forest at the Battle of Kulikova Field in 1380. Seeing a smaller force than they anticipated, the Tartars attacked, whereupon they were overpowered by the concealed troops.[3] During the time of the Tsars and the Bolsheviks the importance of deception was widely accepted within politics but, since both deception and surprise were regarded principally as tools of the offensive and Russia had been on the strategic defensive since the time of Napoleon, the strategy in both 1812 and 1914 was to allow the vastness of the Motherland to swallow an invader. The development of military art in the young Soviet Union was largely evolutionary, but as early as the mid-1920s the Soviets concluded that the 'basic method for the achievement of surprise is operational deception'.[4] They relied greatly on experience, which obviously the Great Patriotic War provided in abundance but which was apparent as early as the Battle of Khalkin-Gol against the Japanese in August 1939. The Soviets under Zhukov drove the Japanese back to Nomohan, largely thanks to their experience of cover and concealment of offensive preparations. Zhukov later described how

> these [deception] measures aimed at creating an impression that we were making no preparations for an offensive operation. We wanted to produce an impression on the Japanese that we were merely building up our defences and nothing else.[5]

Although the Red Army's Field Regulations of 1936 and 1939 stated clearly the need for all types of deception in order to achieve surprise, there was a large discrepancy between theory and practice. In 1941 both the political and military hierarchies believed that the First World War pattern would pertain and that there would a time-lag between the declaration of war and the commencement of operations.[6] Thus the devastating effects that German surprise achieved in 1941 served to focus Soviet minds on both the importance of surprise and the possibilities offered by deception, even though initial chaos and technical incompetence effectively scuppered deception in the early months of the war.

The first operational instruction concerning *maskirovka* was issued on 26 June 1941, just four days after the invasion, and concerned concealment of objectives from the air. During the early war years the Red Army had to learn painfully the art of modern warfare, of which *maskirovka* was but one aspect. It was hampered in this by poor radio discipline, although it learned camouflage techniques and could mask the movement of large forces if radio use could be avoided. Such success as the Soviets did enjoy in the first year of the war was probably due as much to the chaotic nature of the fighting and the rapidity of the German advance as to improvements in Soviet security measures.[7] Thus the first major Soviet attempt at *maskirovka* – to cover preparations for the counter-offensive in front of Moscow in December 1941 – succeeded not so much because of its deceptive effect as because of good security, radio and light discipline, and night movement to help the concealment of regrouping forces; moreover, the Germans completely underestimated the regenerative powers of the Soviets. On 18 November the Chief of Staff of the German Army, Generaloberst Franz Halder, wrote:

> [Generalfeldmarschall Fedor] von Bock shares my deep conviction that the enemy, just as much as we do, is throwing in the last ounce of strength and that victory will go to the side that sticks it out longer. The enemy, too, has nothing left in the rear and his predicament probably is even worse than ours.[8]

The deployment of Zhukov's Western Front (army group) of 1st Shock, 10th and 12th Armies took place in the strictest secrecy in a manner that would become standard practice for the future. Strict light and camouflage discipline was observed and all movement carried out at night under absolute radio silence. Particular attention was paid to disguising supply depots and the rail and road communications along the deployment routes. The Germans, who were preoccupied with their own problems and tended throughout the war to dismiss Soviet abilities, were largely unconcerned. However, Halder noted on 29 November that 'on the front of Fourth Army ... there is some talk that the enemy is preparing to attack'.[9] But Halder otherwise displayed optimism, as did Army Group Centre: nowhere on their maps did the 1st Shock, 10th or 12th Armies appear.[10] Even one whole day after the offensive started on 5 December, OKH had still failed to identify three of the ten armies in the Moscow sector. Within a week German forces were struggling to withdraw in good order from the Moscow region, an indication of what might be achieved with *maskirovka*.

STALINGRAD

In the early period of the Great Patriotic War most deception operations took the form of *imitatsiia*, as for example before an attack by the Western Front across the Lama River west of Moscow in February 1942. In order to lure away German bombers from the real concentration area, the 20th Army built a concentration of hundreds of dummy tanks, vehicles, artillery and troops on its right flank, with simulated firing and broadcasts of the sound of tank motors to add realism. This reportedly diverted 1, 083 enemy flights. During the spring the Germans prepared for a major strategic offensive aimed at the economically valuable Caucasus and supported by a deception, Operation KREML ('Kremlin'), designed to convince the Soviets that the true aim remained Moscow. At the same time the Soviets, with their gaze already transfixed by events in front of the capital, planned a strategic defensive with limited operational offensives. Following one such at Kharkov, Stalin ordered a retreat in the south when the German drive towards Stalingrad began, a combination that was to have fatal consequences for the Germans. It also precipitated a leadership crisis when the retreat threatened to become a rout, and as a result Zhukov was named Deputy Supreme Commander and took *de facto* control of operations. The first result of this appointment was Operation URANUS, which aimed at tying down the German Sixth and Fourth Panzer Armies until November without allowing them to settle into a defensive posture; a counter-stroke would then be unleashed. It was imperative both that the Germans should not capture Stalingrad and that they should remain completely ignorant of Soviet intentions. The *maskirovka* plan would need to be of an unprecedented sophistication.[11]

As in late 1941, so during summer and autumn 1942 the Stavka (Soviet High Command) carefully created reserve armies and also formed new armoured formations. These were mechanized corps and tank armies where previously they had only deployed armour in brigade-size groupings. As the Stavka examined the experiences of the first year and a half of war, they noted that false objectives had been created unsystematically and unrealistically, and had produced no discernible reaction from the enemy. Poor camouflage in rear areas and of roads and the lack of participation by chief engineer officers in formulating operational plans had resulted in poor camouflage security. A directive was issued to rectify these matters, which also specifically demanded greater efforts at disinformation, a directive that became the focal point for the Stalingrad operation.[12]

From these studies recommendations were made for masking offensive preparations, including thorough camouflage and concealment, ensuring all reconnaissance and combat security should be carried out by units previously in contact with the enemy, and movement by night. They had noted that at least one rifle company, three tanks, three anti-aircraft machine-guns and three anti-aircraft guns were normally required to simulate a regimental (brigade) concentration. During the organizing of an offensive every commander was expected to prepare a thorough *maskirovka* plan. It was appreciated that even if the intent could not be hidden, then there was great benefit from hiding the timing. But radio procedure was still not of a high enough standard to permit effective use of false traffic, and improved *maskirovka* discipline within armour, artillery and infantry units could not counteract German air recce. The problem would have to await the defeat of the Luftwaffe.[13] Meanwhile, Operation URANUS would be a test of Soviet ability across the spectrum of the art of war, not least of *maskirovka*.

In August the Soviets launched Operation MARS in front of Moscow. It achieved only limited gains but drew twelve German divisions into Army Group Centre. Some Soviet sources refer to MARS as a deception, possibly modelled on the example provided by the Germans with KREML, which had proved highly successful in persuading the Soviets that the German main effort would be towards Moscow.[14] But MARS was a real operation and rather a disastrous one, although it certainly had the effect of drawing attention towards Moscow. An intelligence summary from Fremde Heeres Ost (FHO, the Eastern Intelligence Branch of OKH) on 29 August concluded that the Soviets would have considerable offensive potential in both Army Group Centre and Army Group 'B's sectors, but that 'the Russians would be more eager to remove the threat to Moscow' and they subsequently forecast a Soviet offensive there. Most significantly, it assumed Soviet inability to make more than one major offensive.[15]

The Stalingrad Front conducted offensive operations between 29 September and 4 October to secure the bridgeheads over the River Don and distract German forces from defensive actions in Stalingrad itself. As a security measure to preserve the concept of operations, Zhukov kept that concept to himself and a few trusted aides and imposed a strict limit on the time allotted for planning to the fronts and armies due to make the actual assault. This was made possible by the Red Army's now improved sequential planning ability. This greatly reduced time framework meant that the front commanders were not permitted to implement their own plans before the first week in November. To 'disinform' the Germans, the fronts were ordered to go onto the defensive from 15 October, so that all visible effort along the frontline was put into building defences. All the villages within twenty-five kilometres were evacuated and ringed with trenches to give enemy air recce something to see. Concealing the direction and scope of the main effort was going to be the hardest aspect and here it was necessary to give way to practical limitations. The need to bring in a third front headquarters – Southwestern – to control the main effort, was delayed until 29 October and generally the build-up was made with formations smaller than army size (with the exception of 5th Tank Army, which would form the spearhead). All movement was conducted at night in strict radio silence and the reserves held 200 kilometres upstream of Stalingrad on the Volga.[16]

The South-Western Front now proceeded to bring 5th Tank Army and 21st Army into bridgeheads south of the Don, north of Stalingrad. For this they built twenty-two bridges (including five false ones), concealing the approaches with vertical covers and camouflaging the crossings. One particularly clever technique was to build the surface of a bridge just below water level or concealed beneath ice.[17] Trucks would get wet tyres but, as the attack date of 19 November drew closer, German aircraft bombed the false bridges and left the real ones intact. Extensive smokescreens were used to cover the movement of 26th Tank Corps into the bridgehead and the engineers constructed simulated concentrations of artillery and tanks to divert German artillery and air recce. The Don Front conducted similar tasks as its commander, General-Major Konstantin Rokossovskiy, recalled:

> much was done in order to deceive the enemy. We tried to convince him that we were about to attack in the area between the rivers and conducted more operations here. In the remaining sectors of the front, intensified operations for the erection of foxholes, fortifications etc., were simulated. Any movement of troops in those regions, from where they were to operate, was carried out only at night, with the observance of all camouflage measures.[18]

The total night movements involved 160,000 men, 10,000 horses, 430 tanks, 14, 000 vehicles and 7, 000 tons of ammunition, and were mostly undetected by the Germans.

Soviet newspapers continued the familiar slogan *ne shagu nazad* ('not a step back!') and talked of offensive plans elsewhere.[19] Finally, Zhukov had to negotiate two aspects of current Soviet doctrine that might alert the enemy to his peril. The first was *razvedka boyem* ('battle reconnaissance'), which was then regarded as indispensable to offensive operations to feel out objectives, defences and the nature of the terrain; it was normally conducted in strength but Zhukov restricted it to battalion-sized groups throughout the Stalingrad area. Second, he ensured the Red Army's Chief of Artillery, Nikolai Voronov, was close at hand to guarantee the secrecy of the artillery deployment and to limit the artillery preparation to an hour and a half.[20]

As a result, while the Germans were aware of offensive preparations they were unaware of the magnitude of what was building against them, a factor exacerbated by optimism and over-confidence. During October, Oberst Reinhard Gehlen, Chief of FHO, forecast an attack in the Moscow area, although Hitler regarded the potential danger to Army Group 'B' as serious. Air recce had revealed some of the bridges across the Don but, significantly, he did not see the danger as particularly imminent. Reinforcements ordered from France would not arrive before December, and in the first week of November Hitler went on leave for two weeks to Berchtesgaden.[21] On 29 October Romanian Third Army reported to Army Group 'B' on the marked increases in crossings of the Don in the Soviet rear and on continuous attacks along the frontline (a reference to *razvedka boyem*) as well as reports from deserters.[22] This prompted Generalleutnant Freidrich Paulus commanding German Sixth Army to report that 'a major enemy attack, it was considered, could be expected in the next few days'.[23] However, Generaloberst Kurt Zeitzler (who had replaced Halder as Chief of the General Staff in September) stated that

> the Russians no longer have any reserves worth mentioning and are not capable of launching a large-scale offensive. In forming any appreciation of enemy intentions, this basic fact must be taken into consideration.[24]

Once again the Germans had fatally underestimated the regenerative capabilities of the Soviets. At 0730 hours on 19 November the order to open fire was given to 3, 500 guns and mortars massed in the Sermafimovich bridgehead. As Romanian Third Army reported several 'weak' attacks and a stronger one at around 0900 hours, German Sixth Army continued to assault Stalingrad throughout the day. At 2200 hours a teletype message arrived at Sixth Army headquarters:

> The development of the situation at Romanian Third Army compels radical measures to secure forces to protect the deep flank of Sixth Army. All offensive operations in Stalingrad are to be halted at once.[25]

It was too late. The Germans suffered a crushing defeat that marked a turning-point in the war, although they managed to retain the strategic initiative until the following summer.

The Soviets, on the other hand, had gained valuable experience in their first crude experiment with strategic deception, having made strenuous efforts to simulate offensive intent elsewhere along the front. German concerns with Moscow and the fighting in Stalingrad itself distracted them from the vulnerability of the flanks of Army Group 'B' and demonstrated the value of those efforts. At the operational and tactical level the rewards were even greater, producing paralysing surprise. The ability to obscure matters of timing, detail and scale would provide the basis for future *maskirovka*.[26]

ADOPTING THE STRATEGIC OFFENSIVE

During the course of 1943 the Soviets' ability to deceive strategically matured with a comprehensive and successful *maskirovka* plan for the Kursk defensive-counter-offensive operations. A series of diversionary attacks along the Mius and northern Donets rivers concealed the intent to attack along the axis Belgorod–Khar'kov–Poltava and drew German reserves away from the chosen sector. Once they reached the River Dnepr they feinted along its entire length while regrouping secretly to the north, and established such a bridgehead that the Germans were overwhelmed. Here on the open steppe where natural cover was sparse, colossal amounts of smoke were used to screen the whole sections of the front, most notably by General-Lieutenant P. Batov's 65th Army during the Dnepr crossing operations. Soviet strategic *maskirovka* in 1943 involved only limited regroupment and concentration of Stavka reserves. The most significant example was the deployment and use of the Steppe Front at Kursk and between fronts on the advance to the Dnepr.[27]

In late September 1943 the Stavka came up with a plan to destroy the German forces between Vitebsk and Gomel in the central sector of the Eastern Front. Within this plan the task of the Kalinin Front was to penetrate the German defences and seize the important city of Nevel, which controlled lateral communications between two German Army Groups. General of the Army Andrei Yeremenko chose to attack on 6 October. He made his main effort from his right flank towards Nevel with the 3rd and 4th Shock Armies, and in order to divert attention from this he ordered diversionary attacks westward towards Vitebsk by 39th and 43rd Armies on his left flank to commence on 2 October. For his plan to succeed it was essential that 3rd Shock Army was able to concentrate secretly and rapidly to launch a sudden attack.

The commander of 3rd Shock Army, General-Lieutenant K. N. Galitsky, in turn planned to attack with two rifle divisions on a four-kilometre front supported on their left flank by two divisions from 4th Shock Army. He would then exploit to Nevel with a third rifle division supported by a tank brigade, retaining a fourth rifle division and two rifle brigades in reserve. For the rest of his hundred-kilometre frontage he deployed two under-strength rifle divisions and a rifle brigade and organized two fortified regions; these in turn would form one or two battalion groups for local attacks to divert attention from the main effort. This aimed to capitalize on the swampy and heavily wooded ground to the east of Nevel, both to mask the preparations and to hinder German counter-measures. Galitsky's plan was approved by Front headquarters on 27 August under the strictest possible security. Only Galitsky and his chief of staff, then subsequently his chiefs of services and operations and two assistants, helped to formulate the plan, preparing a limited number of documents by hand.

Regrouping began on 3 October, closely supervised by Army staff officers along previously prepared routes through the forests and physically covered in key sectors to prevent German air recce spotting the build-up. All movement was strictly timetabled and carried out in radio silence. At the same time 39th Army carried out similar movements in the south in more open fashion to lend credibility to the diversionary attacks towards Vitebsk. The regrouping was completed by 5 October and commanders made reconnaissances of the approach routes to their start lines, which were physically concealed by the engineers. After dark on 5 October the first echelon units moved into their forming-up points, just 300 metres from the German front line. The assault commenced at 1000 hours, whereupon Galitsky committed his second echelon division on the flank where greatest progress was being made. The attack achieved such overwhelming surprise that, with the support of the tank brigade, a penetration of over twenty kilometres was made to capture Nevel by 1530 hours, testimony to the effects of both active and passive *maskirovka*.[28]

Shortly afterwards, 3rd Guards Tank Army completed what was possibly the most profitable use of *imitatsiia* during the war. They pulled out of the Bukrin bridgehead on the River Dnepr south of Kiev and moved into the Lyutezh bridgehead to the north of the city in order to launch a surprise attack that enabled the liberation of the city by 6 November – in time, as Stalin had ordered, for the celebrations of the October Revolution the following day. In the *maskirovka* operation, command radio stations of the formations involved remained in place and the tanks and vehicles moving out of position were replaced with dummies. When the movement timetable slipped, fog cover was enhanced with smoke, which was also used to protect the bridge. Within the Bukrin

bridgehead, radio and artillery firing patterns continued as usual and rumours were promoted of a renewed
offensive to the south. False road traffic was maintained
throughout daylight and mock-up huts and dugouts were
built in the rear areas.[29]

Other operations in 1943 were less successful (some
possibly deliberately). Along the Mius River in July and
along the northern Donets River in July and August
poor radio security undercut the effectiveness of the
maskirovka. In both of these cases the Germans were
able to move reserves to the threatened sector and halt
the Soviet offensives with minor losses. These reserves,
however, were drawn from more important sectors of
the front. Throughout the winter of 1943–4 the Soviets
launched a series of offensives starting in Ukraine, continuing in the north to relieve Leningrad and culminating
once more in Ukraine in March 1944, with a drive that
cleared the Crimea and approached the pre-war borders
of Poland and Romania. All of these operations involved
maskirovka with varying degrees of success. Again, while
it was impossible to conceal offensive intentions, masking of timing and main efforts often brought considerable
gains.

Over the years *maskirovka* had developed from largely
passive efforts to maintain security into more concerted
and co-ordinated schemes that required specific staff
agencies within major headquarters. At the same time the
lessons learned were being incorporated into new regulations; the 1944 Field Regulations opened with a eulogy to
the benefits of surprise before outlining specific measures

for achieving deception and stressing the importance of deception in achieving surprise. It stated:

> *Maskirovka* is a mandatory form of combat sup-
> port for each action and operation. The objectives
> of *maskirovka* are to secure concealment to the
> manœuvre and concentration of troops for the
> purpose of delivering a surprise attack; to mislead
> the enemy relative to our forces, weapons, actions
> and intentions; and thus force him to make an
> incorrect decision.

It specified: concealing real objects from the enemy, changing the external appearance of objects, spreading false rumours, noise discipline and artificial noises, and radio discipline, false radio nets and radio deception, and it stressed the principles of 'naturalness, diversity and continuousness'.[30] A *Manual on Operational Maskirovka* was published later that year to give front and army-level planners basic guidance on formulating their plans and set the framework for the development of operational *maskirovka* into the strategic variety.

OPERATION BAGRATION

The *maskirovka* plan for Operation BAGRATION was the largest and most comprehensive that the Soviets ever attempted and the resulting 400-kilometre advance a testimony to its success. During the spring of 1944 both sides prepared for the Soviet offensive expected once the mud had dried. Once more the problem for the Germans was deciding exactly where the blow would fall. The recent Soviet advance through Ukraine suggested that the main effort would be towards the Balkans, promising political gains and covering favourable ground which would give several possibilities for further developments. The Germans were particularly sensitive to the possible loss of the Romanian oilfields and were well aware of both Romania's and Hungary's wavering commitment to the Axis. But several other possibilities were open to the Soviets, including the direct route to Berlin by way of Belorussia. This was deemed by the Germans to be a more difficult option, however, with poor roads and the Pripet marshes to traverse. An alternative would be an attack from Ukraine towards the Baltic which could then isolate Army Group Centre from the rear. Further attacks were also expected in the Baltic region, but here the nature of the terrain would undoubtedly favour the defender and it was considered a strategically less significant problem.

In choosing the Belorussian option, the Stavka sought to exploit the possibility for deception offered by the others; such a choice would also exploit Hitler's wishful thinking, since an offensive out of northern Ukraine presented the best hope for a German counter-stroke. Its subsequent success was in no small measure due to German fears and misconceptions, upon which the Soviets played handsomely. As Zhukov remarks in his *Memoirs*, 'intelligence reports showed that the German High Command expected us to make the first blow of the summer campaign in Ukraine, not Belorussia.'[31] By care-fully regrouping their forces and time-phasing all offen-sive movements, the Stavka created a strategic *maskirovka* plan that would contribute to convincing both Hitler and OKH that the Soviet summer offensive of 1944 would emanate from northern Ukraine. In fact, the offensive aimed at destroying three German army groups on the central and southern portions of the Eastern Front in five successive, distinct operations beginning with a diver-sionary attack north of Leningrad. It would then feign threats in northern and southern Ukraine, threatening to continue the successes achieved in the winter while making a massive redeployment and reorganization for a blow in Belorussia (BAGRATION), to be swiftly followed by attacks launched successfully from northern and south-ern Ukraine.

Once again security was paramount. Only three people other than Zhukov were aware of the whole plan.[32] The first task for the Stavka was the reorganization of the fronts to conform to offensive requirements. The Belorussian Front was renamed the 1st Belorussian Front and the Western Front was divided into the 2nd and 3rd Belorussian Fronts; these three would co-operate with the 1st Baltic Front in the critical Belorussian operation. Formations such as 5th Guards Tank Army would need to be moved from Ukraine to Belorussia and 2nd Guards and 51st Armies from Crimea, but all of these moves would need to be concealed. Meanwhile, an order was sent to the commander of the 3rd Ukrainian Front to encourage Hitler's belief in the threat from northern Ukraine:

> You are charged with conducting operational *maskirovka* measures for the purpose of misinforming the enemy. It is necessary to show a concentration of eight-nine rifle divisions, reinforced with tanks and artillery, beyond the right flank of the front ... The false region of concentration should be animated, showing the movement and disposition of separate groups of men, vehicles, tanks and guns, and the equipping of the region; anti-aircraft guns should be placed at the locations of tank and artillery mock- ups.[33]

This was duly carried out with dummy equipment and false radio nets and generous air cover that nevertheless allowed occasional enemy flights to record the 'build-up'.

Planning for BAGRATION was complete by 14 May and approved by the Stavka on 20 May. The task was by no means simple, with about one million Soviet troops facing 850,000 Germans – not particularly favourable odds. To create the required odds would need the redeployment of five combined arms armies, two tank and one air army, 1st Polish Army, and five tank, two mechanized and four cavalry corps (some half a million men plus all the associated impedimenta). Each front prepared a *maskirovka* plan in accordance with the overall Stavka plan. At each level this was handled by the absolute minimum number of people, with paperwork restricted ruthlessly. The newspapers talked of defensive arrangements and the political commissars gave suitable instruction to the troops. Along the front line all activities were maintained in their usual routine. False minefields were created and defensive positions improved.

The commander of the 1st Belorussian Front, Rokossovskiy (now General of the Army), recalled the measures taken in his command:

> All headquarters were required to maintain constant air and ground control over the effectiveness with which all activities at the front were concealed from the enemy. He was to see only what *we* wanted him to see. Troops deployed and regrouped under the cover of night, while in the daytime trainloads of dummy tanks and guns travelled from the front to the rear. In many places we built fake crossings and roads. Guns were concentrated on secondary lines, from which they launched artillery attacks and were then removed to the rear: dummies being left there on the firing positions.[34]

A particular problem was *razvedka boyem*. It could not be avoided, so it was planned to overload the enemy with the largest *razvedka boyem* of the war, along a 600-mile frontage including the 2nd and 3rd Baltic Fronts north of the BAGRATION area and 1st Ukrainian Front to the south.[35]

The resulting operation proved to be the greatest German defeat of the war, and yielded greater losses in two weeks than even the two months following the surrounding of Stalingrad. The sudden vacuum created in the centre of the line forced the Germans to shift forces from both north and south just as the Red Army was planning to launch offensive operations in these areas, but how much it can be attributed to successful deception can only be measured against German intelligence reports. On 3 May, the day the Soviets issued their directive on deception, FHO issued its forecast for the summer. It envisaged two possible Soviet offensives: one from south of the Pripet marshes cutting north behind Army Group Centre to the Baltic, and the other driving west through the Balkans. The latter was considered the more likely. Even when some signs of Soviet build-up were detected in June, these were dismissed as 'apparently a deception' and increased attack indicators after 16 June caused no excitement and only routine interest.[36]

In part this was a product of rapidly deteriorating German reconnaissance capabilities. It is also probable that among the information they did receive about the Soviet forces opposite Army Group North Ukraine, as distinct from those in front of Army Group Centre, a great deal was from the large numbers of Ukrainian nationalist partisans operating in that area, which would further serve to confirm their wishful thinking.[37] Added to Soviet deception was 'an almost hypnotic self-induced delusion: the main offensive would come against Army Group North Ukraine because that was where they were ready to meet it'.[38] This is hardly surprising given what had gone before, but it cannot detract from the conclusion that the 'system of operational deceptive measures proved its worth. History has shown that the enemy was profoundly misled concerning our real intentions.'[39]

These offensives serve to demonstrate a different approach, particularly at the strategic level, from that of the Western Allies, a product no doubt of the differences between the continental and maritime contexts. Before URANUS, *maskirovka* amounted to secrecy and conceal-ment with simulation displays. Although MARS was con-ceived as a genuine offensive, its effect – possibly unin-tentional – was to reinforce a misconception and create an ambiguity that improved the effect of URANUS. The *maskirovka* for BAGRATION took this further. The aim was not to create fictional formations in the Western manner, but to present an essentially true picture in a totally dis-torted fashion. This meant that even if the Germans had unmasked part or all of the deception, they would still be faced with two possibilities. Put rather simply, while the Anglo-Americans created one force with which they might strike in any one of a number of places, the Soviet technique involved two or more forces but one strike.[40] Following their enormous success in Belorussia there were subsequent operations elsewhere, but on the Polish sector of the front there now ensued a pause that saw one of the more bizarre deceptions of the war.

Operation SCHERHORN was the creation of a notional group of 2,500 German troops led by Oberstleutnant Heinrich Scherhorn which the Soviets used to iden-tify transmitters and investigate the German command and intelligence system. The German High Command received a message from an alleged network in Moscow saying Scherhorn's men were trapped behind Soviet lines at the Beresino River. From then until Scherhorn's last message on 4 April 1945 the Germans expended considerable effort in men, material, and precious aircraft in a vain attempt to rescue him and his men, including sending two SS groups that never returned. At one point Otto Skorzeny was alerted to create a task force for the mission but by then it was March and too late. Hitler promoted Scherhorn and all the men were mentioned in dispatches, but Scherhorn was sending messages under duress, and his detachment had in fact ceased to exist during the BAGRATION offensive, when Scherhorn and some 200 survivors had surrendered.[41]

THE DRIVE ON BERLIN

When the Soviet offensive was renewed in early 1945, unsurprisingly it was under the cover of a comprehensive *maskirovka* plan. The drive had to come out of the bridgeheads already established over the River Vistula, which limited the possibilities for deception. The 1st Ukrainian Front would attack out of the Sandomierz bridgehead, followed two days later by Zhukov, now commanding the 1st Belorussian Front, from those at Pulavy and Magnushev. Both fronts required substantial reinforcements, and the first aim of the Stavka's *maskirovka* plan sought to conceal this effort; the second aim was to divert German attention towards secondary sectors, particularly on the region south of the Vistula. Employing the security methods so assiduously learned over three and a half years of war, Zhukov created a simulated force concentration on the left of 1st Belorussian Front with 1,000 dummy tanks, self-propelled guns and vehicles while defensive work continued in the attack sectors. All offensive engineer work was carried out at night and senior officers made regular inspections from the air.

In 1st Ukrainian Front's sector the need for effective *maskirovka* was even greater and a simulated build-up was made on the far left of the bridgehead. Soviet engineers constructed thirteen 1-kilometre bridges across the River Vistula. In 8th Guards Army's sector there were three. Two of these entered forests on the far bank and were used to convey traffic into the bridgehead while the third, which terminated in an open region, carried the return traffic. Sixtieth Army organized a command group to implement its *maskirovka* plan. Headed by the deputy chief of operations, it comprised representatives from each branch of service (infantry, armour, artillery, signals, engineers and services) and controlled forces assigned from front assets to perform *maskirovka* tasks. These included an engineer brigade plus two battalions and a separate *maskirovka* company. Sixtieth Army itself provided an engineer battalion and three infantry battalions together with artillery batteries to simulate adjustment fire, officer billeting parties, 200 dummy tanks and officer reconnaissance parties from the defending regiments in the front line. These would make ostentatious recces of forward areas while the billeting parties went around warning the local population of the influx of new units. When 4th Guards Tank Corps moved through the region, it was replaced with over 600 dummy tanks while special detachments animated the area with vehicle movements and camp fires. Roving guns simulated artillery adjustment fire and soon afterwards 550 dummy guns were brought in. German artillery continued to fire at these positions long after the real guns had been moved to the assault sector.

Since German observation posts could see somewhere between five and eight kilometres behind the Soviet front line, the vast reserves that the Stavka allocated were hidden in the small forests scattered throughout the bridgehead by vertical vegetation screens built by the engineers, which extended out into the open areas. No less than 240 kilometres of these masks and 180 kilometres of cross-country roads were built in and around the forests. All of this was backed up by intensive rear-area security measures pursued with increasing effectiveness by the NKVD, which greatly undermined the German intelligence effort. But even with these measures, the Soviets knew that the Germans expected an attack. They therefore determined to try new assault techniques. The assault would begin with a platoon from each lead battalion supported by a couple of tanks, a self-propelled gun and several dummy tanks, which would stage demonstration attacks along the front thirty minutes before H-Hour in order to draw the Germans into their defensive positions before the artillery preparation. Fifth Guards Army also made extensive use of smoke to cover its demonstrations. Similarly, both fronts changed the pattern of their artillery preparations, staggering them with various forward movements.[42]

The Germans certainly expected an attack from late October onwards, but remained uncertain as to exactly when and where it would occur. Their intelligence summaries were in fact fairly accurate regarding these, but were hopelessly wide of the mark concerning the scale of the assault. They missed the deployment of no fewer than six armies from Stavka reserve, so that in all three bridgeheads the Germans thought they faced odds of about 3:1 when in fact they were between 5:1 and 7:1, and even between 8:1 and 16:1 on concentration. If it seems more a failure for German intelligence rather than a success for Soviet *maskirovka*, then it should be noted that by 26 January, when the German front was virtually non-existent in Poland, German intelligence correctly located every first-echelon Soviet army and most corps and divisions. As David Glantz puts it, 'the Soviets masked what they wished to mask.'[43]

If the Battle for Berlin seems anti-climactic compared with what had gone before, the Soviets were also well aware of what had happened in similar circumstances in 1760. They estimated that the Germans still had field forces of a million men and, for all that they were a shadow of the once proud and mighty Wehrmacht, they could still offer credible resistance. The honour of leading the assault fell to Zkukov's 1st Belorussian Front from the Küstrin bridgehead on the River Oder. Zhukov's *maskirovka* plan sought to distract German attention from the central Küstrin bridgehead position by simulating attack preparations north at Guben and south at Stettin while portraying a defensive posture at Küstrin. At this stage of the war, however, *maskirovka* had limited application since the front had narrowed to a few hundred kilometres. Also significant was the production of a huge smokescreen, no less than 310 kilometres long; 92 kilometres to conceal genuine preparations and 218 merely to confuse.[44]

With the defeat of Germany, the Soviets were finally prepared to take a hand against Japan, since Stalin sensed that there could be rich political pickings as the war in the Far East also drew to a close. The scale of the undertaking was vast: the operational theatre was 1,000 kilometres by 600 kilometres; both sides had around 700,000 men deployed in Manchuria and the Soviet far east respectively, a figure that the Soviets would have to raise to double by transporting as many men again along the 9,000 kilometres of the Trans-Siberian railway together with all their arms, equipment and supplies. Assuming that these forces could be assembled secretly, the terrain to be covered in the proposed advance meant that the Japanese would have to be deceived as to the actual routes. Added to this, the campaign would have to achieve its goals in thirty days.

To begin with, the month chosen for the attack was August, which in Manchuria is the month of rains. These could be expected seriously to impede movement and so, even as the Soviet build-up became apparent, the Japanese estimated 1946 as the earliest date for a Soviet attack.[45] It seemed that surprise would also depend to a large extent on the form of the attack, which needed to pre-empt Japanese defences or paralyse the command and control structure, using unusual combat techniques.[46] The main blow, delivered by the Trans-Baikal Front, was therefore delivered across difficult country regarded by the Japanese High Command as impassable for large numbers of troops. At the same time the Japanese did expect an attack towards Hailar and it was deemed important not to disappoint them, lest they should redeploy the large numbers of troops defending it. The 36th Army was therefore detailed to do this, although it mostly comprised low-grade infantry.[47] The success achieved was indeed spectacular and, as the first opportunity for the Soviets to apply *maskirovka* at the outset of a war, would prove an important extension of its theory and make the campaign a special subject for future study.

DEVELOPMENT OF A DOCTRINE

The Soviets learned several very important lessons during the Great Patriotic War. The first was that it is virtually impossible to conceal one's intent to attack, even at the outset of hostilities. However, masking the scale, timing and direction can be at least as effective as concealing the intent: an expectant enemy tends to have a more active imagination and will be more receptive to false indicators, especially if his intelligence service is inefficient. One weakness of Soviet deception planning, though, was its inability to know how successful its own measures were, along with a tendency to follow predictable patterns, especially in the early war years. The Soviets identified concealment of forces and operational concepts as the principal purpose of *maskirovka*, citing the following measures as being fundamental to achieving surprise: secrecy of force deployments; demonstrative actions to deceive the enemy regarding one's actions; simulations to confuse the enemy regarding intent and location of real forces; and disinformation by technical means, false orders or rumour.[48]

Soviet wartime experiences also proved the essential interrelation between tactical, operational and strategic deception measures. Although one could make tactical deception without planning operational and strategic measures, it was impossible to do the opposite. Successful strategic deception depended entirely on the effectiveness of measures at lower levels. Most important was the ability secretly and efficiently to redeploy numerous armies and corps, which depended on the ability to hide individual tanks and vehicles. Sloppy camouflage or radio procedure could jeopardize the whole process, as could over-enthusiastic *razvedka boyem* or artillery registration. It took numerous failures to reveal a talent for *maskirovka*, but by the middle of 1943 that talent was evident. Since it relied on the most extensive application of their methods and techniques, strategic deception took the Soviets longest to master.[49]

Throughout the 1950s and 1960s strategic and operational planning was overshadowed by nuclear weapons, although surprise and deception remained key elements. In 1976, however, General-Lieutenant M. M. Kir'yan, a senior member of the Voroshilov General Staff Academy, wrote that 'surprise is one of the most important principles of the military art', and his list of methods to achieve it began with 'deceiving the enemy concerning one's own intentions'. He further elaborated on among other things, secrecy, camouflage and night movement.[50]

Regardless of its form, the environmental or organizational aspects affecting it, *maskirovka* is governed by four major principles: activity, plausibility, variety and continuity.[51] The first of these principles (*activnost*) states that offensive action is necessary to degrade the enemy's observation capability: his ability to locate and identify troop concentrations and key weapon systems, particularly indicator systems, by the concerted use of electronic warfare, dummies and good camouflage and concealment. Plausibility and persuasiveness (*ubeditel'nyi* and *pravdopodobnyi*) are essential, but their success depends on timeliness (*svoevremennost*). In the large forces available to the USSR there was no need to create entirely false armies, since there were plenty of real ones. Nowadays, far smaller forces are deployed, although the same principles apply. Iraq used a Soviet-based doctrine during the Gulf War in 1991 and, despite deploying over half a million men, made effective use of decoys made of wood, cardboard, paper, cloth and fibreglass, including realistic models of tanks bought from an Italian company.[52] Maskirovka must be varied (*raznoobraznye*), and this requires forethought and originality if it is not to become stale and predictable. It is this embedding of *maskirovka* in the very fabric of every other activity, this level of awareness and training throughout the structure, that perhaps most clearly differentiates *maskirovka* from Western concepts of deception. Finally, continuity (*nepreryvnost*) must be maintained both temporally and throughout all levels of command; a tactical deception error may reveal an operational or even strategic deception.

At the tactical level *maskirovka* includes the following categories: optical/light, thermal, sound, radio and radar.

Optical/light *maskirovka* covers those measures, mainly passive, designed to deny enemy optical reconnaissance systems, including photography. This covers everything from nets, camouflage clothing and special paints to the use of small lights like miners' lamps, worn on the head and pointing downwards so that light can be applied only where needed. But it also includes displays of dummy equipment that are designed to be seen, as thermal *maskirovka* includes both concealing heat sources and creating false ones. Equally, radar *maskirovka* involves methods of reducing signatures, from topographic analysis in order to locate radar dead ground which cannot be scanned, to the application of stealth technology and the widespread use of reflectors to create false radar images. These reflectors (corner, pyramid, spherical or dipole) can also form effective radar jammers. Suspended along a road or throughout an area in pairs, they can mask activity; placed besides a wooden dummy they can give it a radar signature, and they can be used to create false bridges and even to 'alter' the landscape. During the mid-1970s every Soviet motor-rifle battalion was issued with thirty corner reflectors.[53]

From this it would appear that *maskirovka* permeated very aspect of Soviet military life (and by extension, that of modern Russia and other former Soviet states). Indeed, Soviet soldiers were 'compelled by regulations to employ some form of *maskirovka*'.[54] With the threat posed by weapons of mass destruction, this was regarded as absolutely essential, as much to ensure the survivability of Soviet forces as to gain surprise. It was valued primarily for its ability to disrupt and delay the enemy's decision-making cycle and his ability efficiently to target Soviet concentrations and build-ups.[55] Similarly, it is designed implicitly to raise such dilemmas in the opponent's mind as to whether to fire on what may merely be decoys or to accept the risk of a massing of forces close to hand which may later threaten to swamp the defences.[56] Nevertheless, Western analysts were hard pressed when watching Soviet manœuvres to detect the widespread implementation of *maskirovka*. Whether this was proof of its effectiveness, or because the 'real' thing was being held back for operations, or because the practice was far less advanced than history, doctrine and assertion suggested is not clear.[57]

POST-WAR INTERVENTIONS

One Soviet writer noted that

> a more important condition for achieving victory than overall superiority in weapons and manpower is the ability to use concealment in preparing one's main forces for a major strike and use the element of surprise in launching an attack against important enemy targets.[58]

A major theme in post-war Soviet thought was the determination never again to be taken by surprise. In the 1960s and 1970s Soviet military writers began to stress the key role of surprise as one of the important principles of military art. A plethora of articles on the subject culminated in a major work by General S. P. Ivanov, *The Initial Period of the War*, which derived lessons from the events of 1940–41 and August 1945. This need to possess the capability for launching surprise attacks, and to defend against them, became a central theme.[59] The Soviets never distinguished between the tactical, operational and strategic levels of deception, and instead emphasized variation of the means of deception. Among other recognized methods for achieving surprise were the use of exercises and manœuvres as cover for the deployment of forces, a method used in the invasions of both Czechoslovakia and Afghanistan. This was facilitated by the centralization of deception planning in Department D of the KGB's First Main Directorate in 1959, in order to manage high-quality deception operations worldwide.[60]

The invasion of Czechoslovakia in 1968 demonstrated these operations superbly. Contingency planning began several months beforehand, when it was discussed at the highest levels. Although the Soviet Politburo was reluctant to order military intervention, Leonid Brezhnev later admitted that sometime in May they began to contemplate the option as a last resort and began a military build-up, partly as preparation and partly to bring pressure on the reformists to keep events under control. Military exercises also gave cover for the necessary logistic preparations and rehearsals. By late June Soviet divisions had moved from their peacetime garrison locations in Poland and East Germany to the Czechoslovak borders. The first Soviet deployment onto Czechoslovak territory occurred in June and July under cover of 'staff military exercises', following an understanding made between Alexei Kosygin of the USSR and Czechoslovakia's Alexander Dubček, the leader of the 'Prague spring'. Forces from non-Czechoslovak Warsaw Pact countries were not originally to take part in these exercises, and the first units to do so arrived in early June during a meeting of the Czechoslovak Communist Party's Central Committee. They brought with them heavy equipment, including armour and EW assets. They first entered air bases capable of handling the Soviet's heavy lift capability. Not only were Czechoslovak officers not informed of this development, but they were excluded from the post-exercise analysis, a breach of the May agreement about which Dubček complained. It later transpired that the Warsaw Pact command had introduced 16,000 troops into the country between 20 and 30 June. A troop withdrawal announced on 1 July was then delayed until negotiations took place later that month and in early August at Cierna-Bratislava.

These month-long manœuvres formed an unusual deception. They were not only unsealed but widely advertised, and thus served not only as preparation for possible intervention but also to create political pressure. Militarily, they were designed to desensitize the Czechoslovaks and Western leaders and analysts. When it became known on 23 July that the Soviet Politburo was to enter negotiations with the Czechoslovak leadership, the Soviet media announced the holding of the largest logistic exercise ever held by the Soviet ground forces under the Commander-in-Chief Rear Services, General S. Mariakin.

During this exercise, code-named NEMEN, thousands of reservists were called up and civilian transport was requisitioned. The exercise started all over the western USSR and as the negotiations progressed was extended into Poland and East Germany. Immediately before the Cierna conference, major fleet exercises were conducted throughout the Baltic, and all of these exercises continued during the conferences. When NEMEN formally ended on 10 August, a vast air defence exercise began the following day, along with a communications exercise in western Ukraine, Poland and East Germany From 16 August Hungary was included, and the following day the decision to intervene was made.

All this time the KGB was trying to provide 'proof' of counter-revolutionary behaviour to justify military intervention, such as caches of secret weapons 'discovered' near the West German border and fake documents to incriminate the CIA. Czechoslovak stocks of fuel and ammunition had been skilfully reduced by removal to East Germany and the USSR under the pretext of the exercises, and the Soviets arranged for a major exercise of the Czechoslovak Army to take place from 21 August – a day after the intervention was due to start – in order to divert the attention of the Czechoslovak military. Tight security measures were imposed, including radio silence and use of electronic warfare assets, to ensure the West knew as little as possible about what was about to happen. Certainly Dubček himself know nothing until it was too late. Huge forces were deployed, estimated at between a quarter and half a million men, but despite the prolonged logistical exercises the operation was dogged at several points by shortages of fuel and food and water.[61] The Soviets had, however, learned from subjecting the Hungarian Revolution of 1956, when they suffered some 720 dead and missing and 1,540 wounded: in Czechoslovakia they lost only ninety-six men killed.[62]

The invasions of both Czechoslovakia and Afghanistan included the establishment of a military and KGB element to assist in the production of a cover and deception plan to divert attention away from it and allow them quickly to seize the essential facilities and key leaders and officials. In Afghanistan preparations for the Soviet invasion of December 1979 also began months earlier. In April General of the Army Aleksiy Yepishev, head of the Main Political Directorate, led a delegation to assess the situation (as he had previously done in Czechoslovakia). In August General of the Army Ivan Pavlovski (who commanded the Czechoslovak invasion), now Commander-in-Chief of the Soviet Ground Forces, led some sixty officers on a weeks-long reconnaissance tour of Afghanistan.[63] With the country in the throes of civil war following the replacement of the king by Afghan Communists, an exercise held in August involved trans-porting 10,000 troops from the USSR to South Yemen and Ethiopia and back again, in a fleet of Antonov-22 aircraft. In September they took the first steps towards influenc-ing the military situation in Afghanistan during the visit to Moscow of President Nur Mohammed Taraki, and a meeting was arranged with Babrak Karmal, who in due course would adopt the position of president following the invasion. The Soviets were involved in intrigues aimed at eliminating Taraki's rival the vice-president Hafizullah Amin. These backfired and the result was Taraki's death and the ascendance of Amin to power. Forced to accept the coup, they pretended to court Amin and appear to have decided to intervene on a massive scale only as late as November, when they sent the First Deputy Minister of the Interior, General-Lieutenant Viktor Paputin, to Kabul, ostensibly to advise Amin on police and security matters but in reality to rally the supporters of Taraki and Karmal.

Changes to deployments along the Afghan and Iranian borders during this period were apparent to US analysts, who this time were not particularly surprised by the invasion when it came. Preliminary moves began on 8 and 9 December with the lift of airborne units to take control of Bagram airport to reinforce a unit sent originally in September. Their initial task was to secure the main road between Kabul and the Soviet border while other units moved concurrently to take control of Kabul municipal airport. The actual invasion was deliberately timed for 24–26 December, when most Western officials would be on Christmas holiday.[64] On the ground Soviet advisors succeeded in disarming two Afghan divisions by persuading their commanders that they needed to take over their ammunition and anti-tank weapons for inventory and their tank batteries for wintering, and that some of their tanks needed to have a defect modified. Then between 24 and 26 December some 10,000 men of the 105th Guards Airborne Division landed at Kabul while two motor-rifle divisions crossed the border from the north and advanced to take control of key positions in the centre of the country, leaving control of the borders until later. In total, some 80–100,000 men were deployed, and the logistic problems that hampered the invasion of Czechoslovakia were avoided.[65] However, simply taking control of the country's main installations and infrastructure was not sufficient to calm the population and control the country. Although the invasion itself was accomplished with few problems, that was only the beginning.

The invasion was felt by many in the West to be the Soviets' 'Vietnam', and with some justice. Soviet tactics in Afghanistan were very clumsy to begin with, and the poor training of many of the units involved meant there was seemingly little employment of *maskirovka*. As with the Americans in Vietnam, the emphasis was on firepower, using armour and large-scale troop deployments to destroy completely Mujahadeen villages and their associated agriculture. Later, with the introduction of *Spetznaz* (special forces), this changed towards observing arms supply caravans from the air and intercepting them. So Mujahadeen commanded by Abdul Haq took to setting up dummy caravans and assembling a counter-force. Having waited to see where the *Spetznatz* teams were deployed, they would ambush the ambushers. Not many of the Mujahadeen groups were capable of such operations, but only after 1986 did they adopt more subtle tactics.[66]

In the autumn of 1987, during the largest Soviet operation of the war, MAGISTRAL, the 40th Army launched
to drive to clear the main route to Khost district, which
had been effectively cut off by the Mujahadeen. The key
position was the Satukandav pass, thirty kilometres east
of Gardez, and practically the only way through the
mountains between Gardez and Khost. On 28 November,
following unsuccessful negotiations with the guerrillas, General Boris Gromov decided to determine the
enemy's weapon systems (especially air defence) with a
fake parachute drop using twenty dummy parachutists.
This proved highly successful and the guerrillas revealed
their positions for artillery observers to record. They were
then attacked from the air and with a four-hour artillery
programme. Although the deception was very effective,
the artillery programme (which far exceeded Soviet
norms) was not, and the pass was cleared only after heavy
fighting.[67]

Success in guerrilla war is hard to define and body
count is certainly a poor criterion. The Soviets appear
repeatedly to have been engaging rearguards and the slow
or uninformed guerrillas. Night patrols and ambushes
were singular planned missions, not routine events. The
Soviet concept of line-of-communication security appears
to have been to establish a series of fortified positions,
man them and then sit back and wait, without aggressive patrolling or reconnaissance. Similarly, they seem to
have used air power primarily for offensive action and not
reconnaissance, with little effort to shift forces, occupy
temporary sites, or take actions to deceive or 'wrong-foot'

CHAPTER ELEVEN

DECEPTION IN COUNTER-REVOLUTIONARY AND IRREGULAR WARFARE

'It is strange that the commanders of regular forces should so often succeed in small wars in drawing the enemy into action by subterfuge and stratagem.'

C. E. Callwell

SINCE AT LEAST as long ago as 165 BC, when Judah used them against the Seleucids, hit-and-run tactics have been practised by tribesmen, peasants in uprising and even regular soldiers in situations where they were so outnumbered that conventional tactics threatened them with being overwhelmed. The anti-revolutionary uprising of the Chouans in the Vendee and Brittany in the 1790s and the guerrilla campaigns against the French in Spain between 1808 and 1814 and in the Tyrol in 1809 were perhaps the first examples of full-blown hit-and-run campaigns, and all served to demonstrate the savagery and indiscriminate nature of such warfare. But since the last century a confusing array of terms – guerrilla wars, 'brush-fire' wars, small wars, low-intensity wars, counter-revolutionary and internal security operations – has grown up to describe wars between regular armies, usually belonging to a European state, and technologically unsophisticated enemies such as African peoples or North American Indians; but the same labels have also been applied to the partisan and resistance operations of the Second World War and, especially since then, to those campaigns fought by various national liberation movements.

Generally, deception, like the fighting itself, takes place at the small-scale, tactical end of the spectrum. Imperatives of security during the Second World War meant that resistance organizations could play no knowing part in the large-scale deception schemes such as BARCLAY and BODYGUARD, although they could play an indirect role. While the Allies were clearing North Africa and planning for the invasion of Sicily, among the erroneous troop movements that the Germans made was the dispatch of 1st Panzer Division from France to Greece, where it provided targets (and trouble) for the Greek resistance.[1] Resistance activity was encouraged by the Allies during May and June 1943 and was so effective that the Germans sent two more divisions to Greece, including an armoured division. After this had reached the Salonika-Athens line, the Asopus viaduct – the only practical route available for withdrawal – was blown up, closing it for four months.[2] Another way that resisters could contribute was by laying booby traps in areas enemy troops were likely to pass. SOE's camouflage section amused itself in designing exploding cowpats and other deceptive but deadly forms of schoolboy humour. Intelligence reports from resisters were also valuable to deception staffs trying to estimate the effectiveness of their other operations. Used carefully, the pressure these small groups could bring to bear on the enemy at well-chosen moments became an important operational tool.[3]

THE FETTERMAN MASSACRE

In many respects the principles of tactical deception outlined above apply also to irregular and guerrilla warfare. However, there are two recurring features of such warfare worthy of note: the use of ambush and the use of lures to draw the enemy into them. Regular forces will often try to lure guerrillas into an ambush by feigning weakness; and the guerrilla or irregular will often try to lure enemy troops by offering a tempting target for attack, and making up for weakness in numbers and arms by choosing his point of ambush carefully to generate local superiority. Following the American Civil War the American government was keen to expand into the new lands of the west. Inevitably, this brought the settlers sent there into conflict with the plains Indians, whose nomadic lifestyle could not be accommodated in this new arrangement. The arrival of 700 men from 2nd Bn, 18th Infantry, at Fort Laramie with orders to garrison Fort Reno on the Powder River, and to establish two further forts along the Bozeman trail which ran directly through their best hunting grounds, was a provocation the Oglala Sioux could not ignore and they duly prepared for war. The disaster that befell US forces was one that might have been foreseen in the circumstances. The distances between the proposed forts would be a hundred miles, sixty-seven miles and ninety-one miles, too great to be able to provide mutual support or protect traffic, especially when manned only with infantry. Nevertheless, this was what the US commander, Colonel Henry B. Carrington, set out to do. After reaching Fort Reno on 28 June 1866, he continued to the site of Fort Phil Kearny and commenced construction, before continuing on 3 August towards the site of Fort C. F. Smith. His troops were thus already spread too widely to protect travellers, a point he made in a letter to his superior while also requesting a cavalry detachment.

Eventually, sixty men of 2nd Cavalry together with a forty-five reinforcements from 18th Infantry were sent to Kearny, including Captain William Fetterman, who soon became the ringleader of a disloyal clique that regarded Carrington as no more than a political appointee. Among the few not involved in these intrigues was the grizzled senior scout Jim Bridger. 'Your men who fought down south are crazy,' he warned Carrington, 'they don't know anything about fighting Indians.' Few did; but they were learning and Bridger had already established a drill, if attacked, for putting the vulnerable trains carrying wood for the forts into a laager until support arrived. When Fetterman proposed a sweep with a hundred men through the various Sioux encampments along the Tongue River to the north, Carrington told him to go away and come back with a more practical proposal. However, Carrington agreed that more active measures were necessary and sent Fetterman with thirty cavalry not merely to relieve the wood train when it was attacked, but to pursue the Indians along their usual retirement route to Peno Creek, where Carrington would try and cut them off with another thirty-five men. Unfortunately when this happened, as soon as the hundred or so Indians realized they were being pursued by only a third of their number, they turned and counter-attacked. Carrington's party also found itself in trouble and was lucky to escape with only two dead and five wounded.

The Indians on the other hand were delighted to find that their enemies were willing to venture so far from the fort. At 1000 hours on 21 December the wood train set out once more, and an hour later the lookout reported it was under attack. The relief was led by Fetterman with strict instructions not to pursue beyond Lodge Trail Ridge. Fetterman, commanding eighty men (the number that he had once boasted was all he required to ride through the Sioux nation), headed straight along the Bozeman trail and was seen to climb the slopes of Lodge Trail Ridge in skirmish order, engaging groups of Indians as they went beyond the crest. It was around noon. Carrington sent a relief force and at around 1245 hours received a message: 'The valley on the other side of the ridge is filled with Indians who are threatening me. The firing has stopped. No sign of Fetterman's command …'

When Fetterman's defeated force was found, the sight was truly appalling. The bodies were mutilated so viciously that the report was suppressed for twenty years. Only one soldier, bugler Adolph Metzger, who had used his instrument as a weapon until it was battered shapeless, escaped such treatment; a buffalo robe was laid over him as a tribute to his bravery. Although the exact details are unknown, it seems that a promising young warrior called Crazy Horse was prominent in the attack, which involved up to 2,000 braves, skilfully using decoy parties to attack the wood train and demonstrate north of the fort. Once Fetterman was seen taking the route towards Lodge Trail Ridge, the Indians simply led him on, taunting and riding across his front, until the soldiers reached the ambush site.[4]

INTELLIGENCE

In 1896 Sir Charles Callwell published his manual for British Imperial soldiers, *Small Wars*, in which he noted that

> the enemy has no organized intelligence department, no regular corps of spies, and yet he knows perfectly well what is going on.... This arises from the social system in such theatres of war and from the manner in which the inhabitants live. News spreads in a most mysterious fashion.[5]

It was probably less mysterious than Callwell thought, but he had hit on a salient point: intelligence is of enormous significance in irregular and guerrilla warfare.

Since the security forces live within the population, they inevitably become part of the infrastructure. Many individuals – contractors, delivery drivers, local government employees – have access to them, to information about them and their operations. Many of these people may be hostile and even agents of the guerrillas. The barracks and other buildings used by security forces are likely to be easily observed and monitored, providing guerrillas with information on all the routines of both life and operations that take place there. Surveillance of the security forces by guerrillas is therefore a fairly straightforward matter.

The reverse is most definitely not the case, especially if elements of the security forces are in any way alien to the community under their control.[6] These forces are immediately set apart by their uniforms, but are fighting against forces much harder to identify and locate. The 'unconventional' nature of such warfare means that detailed intelligence is of even greater significance. The destruction of the insurgents' infrastructure rather than their armed forces is the key to victory. Moreover, military forces are often expected to operate within the law, which restricts the instances when they can use force. The legal emphasis is therefore on capturing guerrillas and trying them, especially in contexts such as Northern Ireland. In these circumstances deception is as useful as an aid to gathering intelligence as it is for actually combating the guerrillas in the field.

Callwell noted that spreading fictitious information on proposed operations was easy and sure to reach the target. He went on to say that, although troops in small wars found their opponents skilled in the use of ambushes and masters of the art of deception, such opponents were significantly less adept at avoiding such pratfalls than at setting them. Thus, where such traps were laid by regulars, they usually succeeded.[7] Modern guerrillas, it should be noted, however, tend to be more sophisticated.

Small Wars

From the beginning of their campaigns in India, Africa and elsewhere, the British employed intimidation and deception as legitimate weapons. If this seems the antithesis of the 'public school' ethos, then it should be remembered that the empire was built largely by adventurers and free spirits who were often out of place at home, and most of the empire had been won long before the priggishness of later Victorian society established itself. In any case, they justified their use of such tactics by seeing it as a response to the traits of the indigenous communities. They perceived cunning and deceit, for example, as the salient characteristics of the Indian princes, and were quick to appropriate those characteristics for their own purposes, both politically and militarily. Callwell noted:

It is strange that commanders of regular forces so often succeed in small wars in drawing the enemy into action by subterfuge and stratagem. Irregular warriors individually possess the cunning which their mode of life engenders. Their chieftains are subtle and astute. All orientals have inborn love of trickery and deception... history affords numerous examples of such antagonists being lured out of strong positions or enticed into unfavourable situations, by bodies of trained soldiers handled skilfully – so much is this indeed the case that the subject merits a special chapter.[8]

Small Wars is full of examples of how to deceive and trick an enemy, and nobody suggested it was 'poor form'.

Callwell particularly noted the ability of disciplined troops to create an exaggerated impression of the size of a body of troops, especially in attack. Baden-Powell's capture of Wedza's stronghold towards the end of the Matabele operations in 1896 is an excellent example. The stronghold consisted of several kraals perched almost on the crest of a mountain some three miles long, which was joined to a ridge by a neck. While the defenders numbered several hundred, the British force amounted to only 120; the original plan had been for another column to co-operate in the attack but it was unable to do so. Baden-Powell commenced operations by sending twenty-five mounted men to the neck with orders to act as though they were ten times as strong. The guns would bombard the crest while the rest of the force, comprising some hussars, demonstrated against the outer end of the mountain and against the back of it. After some desultory skirmishing the mounted infantry pushed their way up to the point designated, leaving their horses below with seven horse-holders. But the enemy began to assemble in force and seriously threaten the hill party.

Baden-Powell, perceiving their somewhat critical position, sent orders to the guns and hussars to make a diversion. But these had been delayed on the road and were not yet to hand, so he took the seven horse-holders, moved round the back of the mountain and ordered magazine fire, so as to give the idea that there was a considerable attacking force on this side. The ruse was completely successful. The rebels who had been pressing over towards the neck hastily spread themselves all over the mountain, and the arrival of the rest of the troops at this juncture completed the illusion, reinforced when the guns came into action to the front. The hussars moved around the mountain and dispersed to represent as strong a force as possible to impress the enemy It was decided that no assault should be delivered that day; instead the deception was maintained throughout the night. Fires were lit at intervals around much of the mountain and fed by roving patrols. The men had orders to discharge their rifles from time to time at different points, and everything was done to make Wedza and his followers believe that a whole army was arrayed against them. The next day the kraals were captured with ease, after most of the enemy had slipped off into the darkness.[9]

CHINA AND MAO TSE-TUNG

Modern concepts of revolutionary war probably owe more to Mao Tse-tung than anyone else. He understood that 'because guerrilla warfare derives from the masses and is supported by them, it can neither exist nor flourish if it separates itself from their sympathies and co-operation.'[10] At the same time as sustaining the military struggle, this provided the basis for the political development he sought to engineer, namely revolution. Similarly, most modern guerrilla campaigns are likely to be inspired by broader political or nationalist objectives than those of a nineteenth-century local chieftain or ruler trying to preserve his independence. Deception has been a central theme in Chinese literature since at least the time of Sun Tzu, and deceptive skill has been highly prized by the Chinese as a leadership quality. Much of Mao's military writing appears to be based on these traditions, putting stress on knowledge of oneself and one's enemy, using deception to control the dynamic of a situation, to reduce costs and control risks, and the use of deception as a means of helping the opponent to defeat himself. Mao talks of 'luring the enemy in', 'feinting to the east' and 'counter-encirclement'.

From its earliest days the Chinese People's Liberation Army (PLA) adopted the standard guerrilla tactic of attacking an enemy outpost in order to ambush the relief force. The PLA called this tactic *ch'ien niu* ('pulling a stupid cow') and employed it on an increasingly grand scale. Its most effective practitioner was possibly Liu Po-Ch'eng, who commanded the Central China Field Army. Liu used the tactic to good effect in 1946 and 1947 to relieve pressure on the East China Field Army in southern Shantung and northern Kiangsu, effectively diverting Nationalist power into northern Honan. In November 1946 Liu's diversionary offensive against the southern Hopei cities of Shangkuan and Laoanchen drew eight Nationalist divisions in pursuit, which were attacked in ambush and encirclement operations. The Nationalists sustained losses equivalent to a complete division and their weapons were added to the Communist inventory. The tactic was continued into the spring of 1947 with the surrounding of several cities along the P'ing-Han (Peiping–Wuhan) railway north of Cheng-chou in Honan, and destruction of the Nationalist relief columns piecemeal.

In May, Liu was ordered to cross the Yellow River and attack south towards the Yangtze. As the main Nationalist forces were occupied in Manchuria, this would have the effect of transforming the war by threatening the communications to the north and north-west. Liu's crossing operation was covered by diversionary operations in western Honan and Shantung to tie down Nationalist forces to the east and west of the crossing sites, and add to the ambiguity of Communist plans. Shortly before the actual crossing on 30 June, a feint was launched against northern Honan in the west and the crossing made against light opposition. It was, according to Liu, the classical stratagem of 'making a big noise in the west in order to attack in the east'. Liu went on to use another variation of 'pulling the cow' on 14 July when he surrounded three Nationalist divisions east of K'aifeng, leaving an opening through which they tried to escape and ambushing them there. Indeed, the use of diversion and ambush was so widespread throughout the PLA by 1948 that it seems remarkable how easily Nationalist commanders were drawn away from decisive positions into fatal actions.[11]

Pseudo-Operations

Mao was not the only person to appreciate that 'the difference between orthodox wars and wars based on subversion [is] that the instigators of the campaign rely on the people to overthrow the government'.[12] In a number of instances since the Second World War the British have had the chance to learn all about this style of warfare. A significant factor in each case has been the existence of a long-standing political commitment; indeed, the primacy of civil and police authority in such campaigns was made as early as 1934 in *Imperial Policing*, by Major-General Sir Charles W. Gwynn.[13] However, while there was an apparent consensus about how to conduct counter-guerrilla operations from an overall command and control perspective, actual organization and tactics varied widely.

The League of Nations mandate for the administration of Palestine proved a poisoned chalice in 1945, when many of the Jewish defence groups began to attack the British. Numerous Zionist liberation groups sprang up, many of them strengthened by experience gained with the Special Night Squads formed during the 1930s by Orde Wingate. This extraordinary British officer earned the DSO for his exploits in this period, having made extensive use of deception and disguise.[14] The Jewish campaign to establish the state of Israel began on 31 October 1945 and continued for three years. Before Israel came into being at midnight on 14 May 1948, 223 British servicemen had died. The British made little use of deception in this campaign, but the Jewish guerrillas came up with a particularly effective and clever mask aimed at both the British and the Arabs. Whenever they mounted a large-scale attack against the Arabs they wore British uniforms and used British equipment and vehicles, either from stocks issued to them during the Second World War (a Jewish brigade fought alongside the British in Italy) or else stolen in raids or captured from British servicemen. Although the Arabs knew the British had no reason to attack them, they were tricked into attacking British convoys, thinking they were Jews in disguise. As late as February 1948 a raid on a British Army camp near Latrun launched in British uniforms and vehicles resulted in at least five deaths and the capture of large quantities of arms and ammunition.[15]

It was in Palestine that the British first made use of pseudo-operations, in which regular forces disguise themselves as irregulars either to attack and destroy them or (more usually) to gather information in order to direct other forces. Pseudo-operations are best used as a means of gathering long-term and background information, which is normally very difficult to acquire in a guerrilla campaign but essential for attacking the guerrillas' infrastructure. (Tactical considerations – that is, decisions about when and where to attack guerrilla forces – are a different matter.[16]) Some of the Jewish groups opposing the British in Palestine (such as the Stern Gang) were very small and all were very tight-knit and highly organized to the point of impenetrability. Intelligence was very scarce and the police, unable effectively to combat the uprising, were also resentful of senior posts being given to military personnel. Colonel Bernard Fergusson, appointed Commander of the Police Mobile Force, decided to pursue special operations and employed former SAS and SOE people for the purpose. However, the inability to speak Hebrew severely limited their ability to penetrate Jewish groups and poor security further reduced their effectiveness. Two months of largely fruitless operations did, however, point towards lessons for the future. In order to succeed, such operations require attention to the minutest detail, especially racial similarity and language proficiency. Pseudo-operations are extremely difficult to implement in clannish communities or tribal areas and they are also particularly vulnerable to compromise in urban areas. Ideally, all such operations should involve police, service and intelligence organizations.[17]

Malaya was one of the first territories overrun by the Japanese in 1941, an event that helped to shatter the myth of European superiority. During the Japanese occupation the Malayan People's Anti-Japanese Army (MPAJA) provided the core of resistance, although its 10,000 members were largely drawn from the significant Chinese population of the country. The army was led by Ch'in Peng, general secretary of the Malayan Communist Party (MCP), which fully intended to take control of the country in the wake of the Japanese departure. Speedy reoccupation by the British, however, thwarted their plans. Many Chinese former MPAJA guerrillas, formed into ten regiments across the country, returned to the jungle, taking with them the weapons and supplies they had captured from the Japanese. They enjoyed widespread support from the Chinese population if not from the Malay majority. The MCP consisted of a military wing (the Malayan Races

Liberation Army, known to the British as Communist terrorists, or CTs) and a civil wing (the Min Yuen). It was initially estimated that only some 2,300 from a total of 12,500
MCP members were actively involved in military actions,
a total which peaked at 7,292 in 1951 and had dropped to
564 by 1960.[18]

In 1948, after the British had withdrawn from India,
Pakistan and Burma, orders went out to begin a campaign to force the British to leave Malaya and to establish a Communist regime. It was a shadowy war from
the very beginning, only ever referred to by the British
as an 'emergency'. The Chinese insurgents had to try to
persuade the Malays and their own population that they
were going to win and that the ordinary people had much
to gain by supporting them. The British had to persuade
the Malays that the opposite was true and flush out the
guerrillas from the jungle. This was immensely difficult
since in looking for the enemy they advertised their own
presence. The enemy had the advantage of experience and
the concealment that the jungle afforded, and was able to
carry out a series of road ambushes. To begin with, British
command was fractured and the police – who were supposedly the primary anti-guerrilla force – lacked numbers,
especially of Chinese speakers. At the same time, the military intelligence chain was weak, with information lacking proper dissemination.[19] However, once the Director
of Operations, Lieutenant-General Sir Harold Briggs,
introduced the plan that subsequently bore his name in
April 1950, fortified villages and psychological operations proved remarkably effective and eventually ground

down the insurgents. In October 1951 General Sir Gerald Templer became High Commissioner and also succeeded Briggs as Director of Operations. He further streamlined the command and control functions (so that it ultimately took four years to create an efficient structure).[20]

GHQ Far East Land Forces advised the War Office that 'there is scope for deception, not only in the tactical sense in Malaya itself but also in the wider theatre field'.[21] Tactically, one of the most effective deceptions was created by the ability of Malayan Scouts (forerunners of 22nd SAS Regiment) to operate for up to two weeks in the jungle, laying ambushes and taking the fight to the insurgents. When resupply was required, it would be by stealth through porters, while very obvious air drops would be made to suggest that operations were focused elsewhere.[22] Pseudo-operations began with an expanded Special Branch, but the first use of specific pseudo-operations came with the formation of Q-Force Pahang in early 1952 under a Malaya Police lieutenant, Richard Bentham 'Yorky' Dixon.

Dixon pressed for some time to be allowed to use surrendered enemy personnel (SEPs) to form a unit that would impersonate CT units for the purpose of gathering information. At the end of the year he had formed two platoons, but he was killed in action on 20 December. Command passed to Lieutenant Noel Dudgeon and the scheme, which by now had proved its worth, was expanded to other parts of the country before coming into official existence in May 1953 as the Special Operational Volunteer Force (SOVF).[23] However, the effectiveness of the force was limited by the tightness of the Chinese community and by the end of 1953 other policies were proving more successful in isolating the guerrillas from their support and supply sources. By 1955, although the war was not yet over, the guerrilla infrastructure had been greatly weakened and pseudo-operations proved of less value.

Where pseudo-operations were probably most successful was during the Mau Mau Rebellion in Kenya, which took place at the same time as the Malayan Emergency. The central issue was land rights in the 'White Highlands', settled by European farmers on land claimed by the Kikuyu people. The issue was raised by Kikuyu nationalist parties formed during the Second World War and by Jomo Kenyatta's Kenyan African Union (KAU), but there was considerable African distrust of the Kikuyu and the rebellion never spread to the wider population.[24]

Following a series of murders, a state of emergency was declared on 21 October 1952. General Sir George Erskine was appointed Commander-in-Chief East Africa in May 1953 and, besides instituting a series of operations along the lines of those being carried out in Malaya, sought overall command of both military and civil administration such as Templer enjoyed there. But Erskine harboured a deep mistrust of the white settlers and there was much mutual antipathy. His operational priorities were to secure the reserves and make them safe, clear the insurgents out of Nairobi and pursue the 'gangs' (as the guerrilla groups were known) into the forests. This had basically been achieved by the middle of 1954. Thereafter it was a case of concentrating and breaking up or destroying the now isolated gangs in the Aberdare and Mount Kenya reserves and forest areas. There were by now some 4,500 rebels in gangs ranging from just four to five men to several hundred, usually relying on individual leadership and lacking central control. It was this isolation that made them particularly susceptible to pseudo-operations.

Precisely who was responsible for initiating these is open to debate but, although tactical deception was probably tried earlier, the credit seems due mainly to Frank Kitson, then a captain in the Rifle Brigade, who arrived in Kenya in August 1953 with no previous experience of Africa or intelligence work. To improve military intelligence, Erskine appointed army officers to Special Branch as District Military Intelligence Officers, and these in turn controlled non-commissioned officers appointed as Field Intelligence Assistants (later Field Intelligence Officers, or FIOs). He gave permission to Kitson, who was DMIO for Kiambu district, to turn ex-gangsters into pseudo-gangsters after Kitson discovered that one of his captives was remarkably willing to return and act as an informer on his former comrades.[25] As with the SOVF in Malaya, former gang members made the best counter-gang members, although this did not prevent Kitson's enthusiastic assistant Eric Holyoak from joining in. On one occasion Holyoak and the turned gangster James, who was certainly not short of confidence in his playacting, got mixed up with a real gang. James' quick thinking saved the day. James told the gang that he was part of a bodyguard for an Asian who was one of the most senior of all Mau Mau leaders and then introduced

Holyoak. 'How on earth did you pass yourself off as a Mau Mau leader for ten minutes?' Kitson later asked Holyoak incredulously. 'After all, even if they swallowed the story of your being an Asian, there aren't many of them around six foot tall with fair hair and blue eyes.' 'I don't suppose they noticed that', replied Holyoak, 'because it was dark and I was wearing a hat.'[26] Holyoak later joined other gangs with a blacked-up face and passed himself off as an African, a very risky business that required complete faith in the other members of the group, on whom he would have to rely to do the talking.

In May 1954 Kitson was faced with official hostility to his methods and it took intervention from Erskine to set up the Special Methods Training Centre in June to expand the scheme. To be successful, the counter-gangs had to resemble the real ones exactly, so that later Kitson even allowed wives and girlfriends to accompany gang members, as the real gangs did. Kitson's emphasis was very much on developing background information about the infrastructure, while others, notably the police, used the same methods to eliminate the gangs themselves. The Kenya Regiment (formed exclusively from white settlers) also ran pseudo-operations, which Kitson felt ran the risk of compromise, but they proved highly successful and the two forces cooperated to a degree. By 1955 pseudo-operations were very well established, with offensive operations taking increasing priority.[27] The remaining gangs were naturally so wary that only extremely sophisticated counter-gangs could succeed and Erskine's successor, Lieutenant-General Sir Gerald Lathbury, relied principally on Superintendent Ian Henderson of Special Branch (a Kenyan who spoke fluent Kikuyu) in his efforts to subdue the last pockets of resistance in the Aberdares. The Army withdrew from operations in mid-November 1956, leaving the police in charge, and the Emergency officially ended in January 1960.

In no other campaign have pseudo-operations proved so successful, largely because communications had broken down within the Mau Mau organization. It was the right tactic at the right time, incorporated by Erskine into the overall strategy.[28] During the Rhodesian War of Independence, which began in 1966, the government also made widespread use of pseudo-operations. The Selous Scouts were formed specifically for this task, which they carried out between 1973 and 1976. Thereafter, a breakdown in the intelligence process led to their employment principally on external raids, a significant change of role.[29]

During the Cyprus Emergency (1959–64) and in Aden (1964–7) pseudo-operations were attempted with little success. In both cases the guerrillas were tightly knit and clannish, which made penetration very difficult.[30] In Aden the SAS used Fijians and others of similar appearance disguised as local Arabs in 'keeni-meeni' operations (the name comes from a Swahili term describing the unnoticeable winding of a snake in long grass, a euphemism for undercover work).[31] Some infantry units also adopted such tactics: the recce platoon of 3rd Bn, Royal Anglian Regiment, formed a ten-man group that was responsible for capturing fourteen guerrillas including the second-in-command of the Front for the Liberation of

technological superiority. Still fresh in its memory was the legacy of Korea, where UN troops spent almost two and a half years entrenched on hilltop positions in all-round defence, relying on superior firepower to deal with mass assaults. This was in itself a product of the Second World War: American minor tactical skills were poor compared with those of the Germans, and in both the European and Pacific theatres the Americans used overwhelming material power to crush opposition and reduce casualties. Given that the US Army expanded in four years from little more than 120,000 men to around 12½ million, lavishly equipped and supplied with every item the nation's colossal production capacity could provide, this was hardly surprising. However, there was already a tendency to disparage stealth, subterfuge and subtlety.

Vietnam was not America's first involvement with guerrilla campaigns. In the Philippines (1901), Haiti (1915–34) and again in the Philippines during the Huk Rebellion (1946–55) US forces were involved directly or in an advisory role. During the Huk Rebellion America provided military advice which helped the Filipinos conduct a highly successful counter-insurgency campaign.[35] The campaign was a model example from the military point of view but, the US military failed to draw on it in Vietnam, even though their involvement there began almost directly after the end of Huk.[36]

In Vietnam a plethora of military, governmental and private bureaucracies and agencies worked with little or no coordination or co-operation. Unlike in the Philippines, where local forces were organized and trained to fight against guerrillas, in Vietnam the US military sought to impose a conventional structure on the Army of the Republic of Vietnam (ARVN).[37] This was because the threat was perceived as being similar to that in Korea of an invasion by conventional forces from the north across the De-Militarized Zone (DMZ). Furthermore, although the war developed along unconventional lines, American participation was wholly dominated, according to Andrew Krepinevich, by the 'Army Concept' of war, which totally lacked

> emphasis on light infantry formations, not heavy divisions; on firepower restraints, not widespread application; on the resolution of political and social problems within the nation targeted by insurgents, not closing with and destroying field forces.[38]

When Special Forces and other advisers became involved they were more realistic about how to approach guerrilla warfare. In 1962 the Special Warfare Division published the *Counter-Insurgency Operations Handbook*, which included pseudo-operations among its tactical methods. However, these were to operate as 'hunter–killer' groups to 'hunt down and destroy elements of local guerrilla terrorist armed bands' rather than long-term information gathering.[39] Besides, the tactics and structure proposed were never accepted by the Army as a whole since it did not coincide with the 'Army Concept'. Instead, the Army relied on cordon and search type operations and the application of overwhelming firepower.

The US Marine Corps was less hidebound and had a long history of unconventional operations. However, after the deployment of US ground forces in 1965 the Marines' area of responsibility was I Corps adjacent to the DMZ, where they were most likely to encounter North Vietnamese Army (NVA) units in comparatively conventional circumstances. Nevertheless, they did institute a number of 'Greek' projects (DELTA, OMEGA and SIGMA) involving long-range reconnaissance patrols to the borders with Laos, Cambodia and North Vietnam. These used 'Roadrunner' teams dressed and equipped to resemble Vietcong to follow the trails and gather information.[40] However, the period when the Ho Chi Minh Trail was working at optimum efficiency in 1970 coincided with the scaling down of US involvement, and Project DELTA was wound up on 31 July.[41] Besides, efforts by the Marines to use more subtle and appropriate tactics in pursuit of pacification ran foul of the commander of US forces in Vietnam, General William C. Westmoreland, who ordered them to take a more aggressive stance.[42] The Marines tried to keep the pacification strategy going and this led to the development of the Combined Actions Platoon concept and the Kit Carson Scouts, who often dressed as Vietcong and preceded Marine patrols. However, such deceptions were only ever used tactically and on *ad hoc* basis. [43]

In Vietnam the Beach Jumpers were more involved with Psyops than with deception. Their commander, Lieutenant-Commander Charles R. Witherspoon, later recalled that battlefield commanders were reluctant to use or condone deception of a more bizarre nature if they could not see an immediate advantage of such an action. Deceptive tactics such as simulating or hiding large-scale force movements were proposed but then rejected because of the extreme physical difficulty of implementation under existing circumstances, e.g. harsh terrain.

At higher command levels efforts to formulate broad deception strategy were hampered by the lack of a unified command structure and political factors; these were exacerbated by the constant rotation of personnel, which contributed to a general lack of the knowledge and co-ordination necessary for sophisticated deceptions.[44]

The rotation of personnel proved to be a serious handicap to the performance of US forces in Vietnam. Writing with hindsight, one American officer commented that his country did not acquire ten years experience but 'one year's experience ten times'.[45] At the same time a further pernicious influence was increasingly apparent. From the beginning of the great expansion of the US Army, the support arms and technical services took the best recruits, and the infantry was regarded as little more than a repository for the dross. Few wanted to join the infantry, especially the type of recruit who would make a good junior leader – the essential requirement in unconventional warfare. With an infantry force lacking the motivation, training or initiative for the close-quarter work involved in fighting an elusive enemy, and with a High Command whose faith in firepower had thus far been repaid, the concept of 'fire bases' was established. The enemy would be crushed with an overwhelming weight of metal, high explosive and petroleum jelly. The B-52 (a strategic bomber), the DC-3 Dakota armed with Gatling guns and the ubiquitous helicopter gunship all took leading roles. Deception, already a forgotten art, was viewed as underhand and un-American. Americans, 'raised in a culture which seeks direct solutions to problems and which hungers after rectilinear forms in work, in play and in battles,' saw deception as just another 'commie' trick.[46] A notable exception was 4th Bn, 39th Infantry, which in 1969 employed a 'special action force' of twelve 'Vietnamese-sized' Americans and six Vietnamese scouts (who were all former Vietcong) wearing black pyjamas and equipped with AK-47s and other captured equipment, for covert and deception operations. But this was very unusual.[47] Thus, although some US forces used minor tactical deception to a modest degree, Military Assistance Command Vietnam failed to develop or co-ordinate it. The Vietnamese, by contrast, used it to devastating effect, both tactically and at the operational level.

A favourite tactic was to ambush a small South Vietnamese or US unit in order to lure larger relief forces into a bigger ambush. A typical example came on 21 February 1966. The Vietcong ambushed a district chief on a road north of Plan Thiet. The 88th Regional Force Battalion rushed to the scene and was also ambushed. Three battalions of the regular South Vietnamese Army (ARVN) were sent to deal with the incident and this time chose to approach from a different direction. They were able to catch 602nd Vietcong Battalion preparing a massive ambush and force them to withdraw. Although government forces eventually did learn to deal with this threat, they seldom took the initiative themselves.[48]

There has been much discussion about deception within the soul-searching that has accompanied American involvement in Vietnam, and in particular about the Communist use of it on the political front. A major weakness of American policy was its failure to recognize the synthesis of all efforts, including political ones, by their enemy. Although in guerrilla warfare the military cannot be separated from the political, the American military tried to do precisely this, most importantly in the field of intelligence.[49] Both sides recognized the doubts that lurked under the surface of American public opinion and saw them as constituting the West's Achilles heel. Between 1965 and 1967 the Communists succeeded in portraying the National Liberation Front as an independent and indigenous southern political entity with a policy of its own, fighting for the cause of freedom, independence and justice, and this coincided with the independent judgements of those in the West who blamed the war on President Diem and knew the shortcomings of his regime. The combined effect was erosion of domestic and international support for the policy of the US in Vietnam, because the legitimacy of the policy was doubtful.[50]

The increasing US involvement could hardly have been expected to generate any less interest than it did, certainly considering the increasing American casualties suffered during 1967. However, the influx of men and material had severely weakened the Communists and during the summer of that year there was considerable debate in the North as to how to proceed. The debate was eventually won by the hawks, led by General Vo Nguyen Giap, who proposed a spectacular all-out offensive throughout South Vietnam using every available Vietcong and NVA asset, which he hoped would provoke the longed-for general uprising of the South's oppressed masses. This presented a nasty dilemma: how to issue an order for such an assault and yet preserve the security necessary to achieve surprise, without which such an offensive was doomed. The deception plan that resulted cleverly took into account what the USA expected and indeed actually wanted to happen, combined with a mixture of contradictory signals reinforcing known American prejudices in a blend of what the North termed passive and active measures.

Supplying the cadres in the cities would take time. Weapons had to be brought down the Ho Chi Minh Trail and smuggled into the cities by agents, often women and children, using many different ruses: concealed under loads of produce, for example, or in bogus funeral processions. Passive measures in the diplomatic and political sphere were designed to suggest moves towards a settlement during 1968. In October 1967 it was made known that the Vietcong would observe a whole week of ceasefire during the Tet Festival, which began on 31 January. The decision to attack at Tet was a controversial one but Giap reasoned it would provide perfect cover, and besides there was a precedent: similar attacks had been made against the Chinese in 1789.⁵¹ Although it was a long truce, given that one of some sort had been observed every year for twenty years, it was expected and welcomed by the Allies. Significantly, US analysts regarded it as a sign of Communist weakness and an indication that victory was near: they expected (and hoped) to see the North putting out feelers for peace. When a Vietcong agent was captured by the ARVN and said that he had been sent to open a channel for negotiations with the Americans to discuss matters such as prisoners of war, US authorities tried to force some of his demands on early release of Vietcong prisoners on a reluctant South Vietnamese government. This played on the neo-colonialist fault line that the North perceived as a weakness between their enemies and also served to raise American hopes. The strategy was further reinforced by the North's foreign minister offering substantive peace talks if the Americans ceased bombing them.

At the same time Giap took active military measures. He planned to draw American attention away from the general uprising by offering two alternatives, both near the border and therefore a long way from the true targets in the heavily populated areas around logistics bases. An attack was launched at Dak To on 4 November 1967 that lasted until the end of the month, and another was planned to begin against the base at Khe Sanh ten days before the main offensive.⁵² Nevertheless, there were ample indications of what was impending if only these were noted. Inevitably reports reached the ears of US officials, and a press release was even issued on 5 January 1968 regarding some captured documents which stated clearly that the People's Army was to 'use very strong military attacks in co-ordination with the uprisings of the local population to take over the towns and cities'. Unfortunately, few if any believed the threat was genuine. One reporter picked up a copy of the press release and wrote a single comment beside the translated attack order: 'Moonshine'.[53] And he was far from being the only sceptic. American officers were certain (and were ultimately proved correct) that the Communists were incapable of seizing and holding the major cities. Alas, capabilities do not necessarily coincide with intentions, and the very boldness of the plan made it seem incredible. Besides, if such reports were true it would seriously upset the routine and climate of comfort and convenience that had existed hitherto. In addition, the fact that the plan made a mockery of all the reports of impending success meant that it was dismissed by strongly entrenched interests, notably political ones, on the American side. In fact, the US intelligence system was so fractured and beset with infighting that the Communists hardly needed a deception plan. As some embittered US intelligence officers put it:

The NVA would have done better *not* to have tried to attack the US Military Headquarters in Saigon on the night of Tet; success would only have *ended* the existing confusion within the ranks of the US Command in Vietnam.[54]

The attack on Khe Sanh was highly significant as a diversion from the impending offensive. An attack of some sort was expected, probably against an isolated garrison and aiming for a decisive victory along the lines of Dien Bien Phu; all the indications, therefore, were that the attack on Khe Sanh was the big one. It prompted Westmoreland to issue a warning that 'attempts would be made elsewhere in South Vietnam to divert and disperse US strength away from the real attack ... at Khe Sanh.' Yet there continued to be ample indicators that something far more ambitious was afoot. When the Tet ceasefire came into force on 29 January, it was compromised by a number of cadres attacking prematurely, which might have severely threatened its chances of success. Some local measures were taken that helped to blunt the effect of the attack when it came, but generally warnings were issued too late or were not fully heeded. When it came, the offensive achieved devastating surprise.[55]

Over half a million American troops were more than sufficient to see it off but given that both the American military and government had been assuring the public that the war was all but won in late 1967, the public reaction to Tet, particularly to the small incursion by fewer than twenty Vietcong guerrillas into the American embassy in Saigon, could hardly be blamed on Communist propaganda. That Tet was a resounding military defeat for the Communists could in no way make up for the US military's loss of precious credibility with the American public. It may not have been quite what Giap and the North Vietnamese intended, but the result was the strategic defeat of the world's most powerful nation.

NORTHERN IRELAND

In Northern Ireland many deceptions have been practised by both terrorists and security forces. Although the troubles started in August 1969, it was not until 5 February 1971 that the first British soldier was killed. The following month three off-duty soldiers, including two brothers (John and Joseph McCaig, aged seventeen and eighteen) were drinking in a bar in Belfast when they met three IRA men, including Paddy McAdorey, a prominent member of the organization. One of the IRA men had served in the British Army so conversation was easy, and after going to another pub it was suggested that they go to a party to find some girls. Crammed into cars, the group headed west through the city up the hill towards Ligoniel. At a deserted stretch of the road they stopped and the three soldiers climbed out to relieve themselves. As they did so, they were shot in the back of the head. The brothers were found slumped over each other and their friend Douglas McCaughey was still clutching his beer glass. It was the beginning of a rapid spiral into full-blown and merciless violence. Widespread revulsion at the nature of the killings led the IRA to issue a denial but it was not long before they tried the tactic again. In March 1973 four NCOs were persuaded by some girls to go to a 'party' in the Antrim Road, where three were shot dead and the fourth seriously wounded.[56]

From the beginning it was difficult for the security forces to gain accurate information from the tight-knit and highly localized Republican communities, especially as there was mutual distrust between the Army and the

Royal Ulster Constabulary (RUC). In December 1970 Frank Kitson, now a brigadier, was commanding 39th Brigade in Belfast. The following spring, in conjunction with the Commander Land Forces, Major-General Sir Anthony Farrar-Hockley, he set up the Mobile Reaction Force (MRF). Although its role has subsequently been sensationalized, the purpose of this unit was limited to surveillance and information gathering, usually by two- or three-man teams in suitably battered cars who observed potential bomb targets and photographed suspected 'players', as IRA men were known.[57]

After Kitson's departure in April 1972 the MRF became associated with more unorthodox deceptions. When General Sir Frank King took over command of the British Army in Northern Ireland, he decided to concentrate on Belfast. He believed in providing the soldiers on the ground with detailed intelligence about the daily lives and habits of the players and understood the need to concentrate on gathering information, however trivial, about their routines and activities. Much of the surveillance continued to be carried on by plain clothes soldiers in cars, but other operations involved establishing observation posts in attics, running phoney businesses (including a massage parlour) and sending women soldiers around as door-to-door cosmetics saleswomen. One of the most ambitious operations was the Four Square Laundry. This allowed soldiers not only to observe Catholic areas closely but also to collect dirty clothes and look for traces of firearms and explosives, as well as following the movements of IRA members. Many of these operations had been compromised by October 1972, but it was the court proceedings in February and June 1973 resulting from a shooting incident on 22 June 1972 that blew the MRF's cover. The force was subsequently transformed into the 14th Intelligence Company, however, and proved very effective.[58]

Meanwhile the Army information service set about deliberately disrupting IRA morale and trying to break up its internal cohesion by releasing false information. It frequently announced, for example, that an arrest had been made on the basis of a tip-off from an informer when this was not the case. Such announcements led to a spate of punishment shootings as the Provos dealt out their rough justice. The introduction of trial without jury and a reduction of the burden of proof meant that IRA members were under increasing threat of imprisonment. Between April 1973 and April 1974, 1,292 people were charged with terrorist offences.[59] A hard-hitting report by Amnesty International in 1978 described how easy it was for the security forces to obtain convictions by extracting false confessions. In this atmosphere there was enormous pressure to turn informer, even though discovery meant almost certain brutal death at the hands of the terrorists.

In May 1974 two such young men, Vincent Heatherington (aged eighteen) and Myles Vincent McGrogan (nineteen) were arrested separately along with three other youths in connection with the murder of two policemen. The others were released without charge but Heatherington and McGrogan were sent on remand to Crumlin Road Gaol, where they were offered the choice of being held in the IRA wing or the non-political wing. They chose the IRA wing, but the IRA commander was suspicious and they were interrogated. Heatherington soon admitted that he had been told to enter the IRA wing of the prison by British plain clothes soldiers after being threatened with being linked to the murder of the policemen. Soon after, the two youths were moved from Crumlin and were tried, and acquitted, of charges relating to the murders of the policemen. Although they disappeared once more, they were tracked down by the IRA and shot.

The incident led to a witch-hunt among IRA prisoners, involving interrogation methods such as piano wire and electric current, which forced some to admit to being turncoats when they were not. The result was a killing spree that lasted for a year and claimed a number of prominent Belfast Republicans. Only much later did the IRA discover that Heatherington had not broken under pressure, but had released 'programmed' information designed to tear them apart from within. One Republican later admitted:

It was very much in the mould of the MRF operations, only better planned and it must be said, brilliantly executed. It created paranoia in the ranks and the IRA found it difficult to admit that British Military Intelligence was so good. It almost destroyed us.[60]

Chapter Twelve

THE FUTURE OF DECEPTION

'We must be ready to employ trickery, deceit ... withholding and concealing the truth.'

V. I. Lenin

'I consider it essential that the US forces should continue to take those steps necessary to keep alive the arts of ... cover and deception.'

Dwight D. Eisenhower

MEDIA OPERATIONS

SUCCESSFUL MEDIA OPERATIONS, like Psyops, deal in truth, and would seem to have little in common with deception. Nevertheless, information management is an increasingly complex business and public information officers, detailed to liaise with and assist the press, perform a very important staff function. Freedom of the press is fundamentally far too important ever to justify imposing military controls, and NATO forces have learned to implement media policies accordingly. While they aim to be as helpful as reasonably possible, their operations must naturally be co-ordinated with the rest of the staff since they cannot afford to give away vital information about forthcoming operations, whether real or deceptive.

It is also important to distinguish between Psyops, which deal with forces, populations or groups by working beyond the established media, and media operations, which deal with the plethora of print and broadcast journalists that attend any conflict. In the age of global communications, media operations – dealings with independent television and news organizations – are increasingly regarded by Western armed forces as vital. The media played a crucial part in the American defeat in Vietnam, where the Communists appear to have appreciated its importance very early on and exploited it very effectively; it has even been alleged that the media actually lost the war for America. Such simplistic exaggeration is not helpful, however. The press cannot be blamed for claims by the government and military during 1967 that the war was all but won, and it therefore cannot reasonably be blamed for the subsequent surprise and shock that Tet produced in the American population.

Nevertheless, there is a tendency towards 'spin' that sometimes appears to come from political motivations rather than strictly military ones. In 1999 NATO bombed a train that was crossing a bridge at Gurdulice in Serbia. The bridge was the real target and it was a tragic episode, yet the film that NATO showed of the incident was not quite what it seemed. It showed the train coming rapidly into view and into the bomb sight, too quickly for the pilot to abort; but for an electric commuter train it seemed to be moving rather quickly. In fact, the film had been speeded up but the reporters took it at face value.[1] However, such a story does NATO far more harm than might have arisen from the admission of a mistake. The British government, worried by adverse reaction to the bombing, had dispatched its own chief spin doctor, Alastair Campbell, to advise NATO on media operations. It could be that this was an example of his handiwork, but whether it emanated from a military spokesman or a civilian, the result was that NATO's credibility was affected. The increasing intrusion of politics into military affairs is something that senior military commanders must consider.

This is a new development. As recently at the Falklands Conflict, although television cameras accompanied the Task Force, the Ministry of Defence was able to exert a strong element of control on everything the media did, because the islands were so isolated. When deception was employed, it was only as an afterthought, and according to one participant it was largely 'schoolboy stuff'.[2] The Argentines made dummy craters out of earth and stones to give the impression that attempts by the RAF to put Port Stanley airfield out of action were more successful than they were. In fact, the Argentines were able to use the airfield by day and night until the time of their surrender.[3] Just nine years later, however, a very different war was fought in the full glare of a global media now in the realm of instant satellite television broadcasts. The public could watch the anti-aircraft tracer arcing across the Baghdad skyline as it happened. The Coalition showed footage of precision attacks on pinpoint targets, and fully in the face of the cameras an operational deception on a scale not seen for nearly fifty years was created.

THE GULF WAR, 1991

After their invasion of Kuwait in August 1990 the Iraqis deployed their forces so that those defending Kuwait would also form the forward defence of Iraq. The key positions were Kuwait City and Basra; the Iraqis assumed no attack could be made through the desert interior, since the major roads were near to the coast and because the interior possessed some very difficult terrain, including rocky and sandy areas almost impassable to vehicles. They therefore concentrated their defences in three lines: on the coast road, the direct route between Saudi Arabian cities and Kuwait; on the bend in the border with Kuwait; and along the road parallel to the Iraqi pipeline to Saudi Arabia, west of the Wadi Al Batin, which formed the most direct route to Basra from the Saudi Army base at King Khalid Military City. The recent war with Iran had led the Iraqis to believe that air power was largely ineffective in support of ground operations and was important only as a 'force in being', held as a strategic counterweight to threaten the enemy, and of no value once aircraft were lost in action. They saw the most important factor as massive ground forces, particularly if placed in dense fortifications with strong artillery support for the breaking up of enemy attacks. They therefore tried to replicate the defensive network that had proved effective against the Iranian infantry during the Iran–Iraq War.

This overlooked the real differences in capability possessed by the Coalition. The Iraqi High Command believed the Coalition would take the direct route towards Kuwait City and Basra, so the defence was layered along the Saudi border and the coast with an infantry 'crust' and armoured units in reserve, while increasingly better formations were deployed further north, culminating in the Republican Guard. Massive firepower could be brought to bear and there was a thick obstacle belt (this tactic was based largely on Soviet doctrine) running from the coast to the Wadi Al Batin. From the start the Iraqi commanders had assumed that there would be an amphibious landing from the Gulf into Kuwait. By a process of elimination they concluded that this would most likely come to the north-east of Kuwait City, whence it would be possible to get access to the north-south highway to Basra and to cut the Iraqi Army in half.[4] Even the defence of the city itself presumed that the main attack would come from the sea. Buildings facing the shore were evacuated and turned into fighting positions and the trench line extended throughout the city along the beach. In total, four armoured divisions and seven infantry divisions were aligned to cover this threat from the sea.[5]

Against an enemy geared to fast-moving warfare and enjoying air supremacy, the Iraqi forces were always going to find it hard to control the direction and tempo of the battle. These problems were compounded by the Iraqi plan, which was heavily weighted to the east, leaving the west vulnerable. There was no proper intelligence operation to assess the Coalition's likely strategy – the surveillance aircraft used in the war with Iran were too vulnerable – only some monitoring of commercial broadcasts and occasional patrols, and this rendered the Iraqis an easy target for deception. Most Iraqi soldiers had little idea of where their own forces were, never mind those of the enemy.[6] Commanding the Coalition during the subsequent Operation DESERT STORM, General H. Norman Schwarzkopf, convinced that newspaper and television reports had become Iraq's best source of intelligence, imposed strict security measures on his own side. Later, when an issue of *Newsweek* appeared to show his flanking plan in precise detail, Schwarzkopf called the Chairman of the Joint Chiefs of Staff, General Colin Powell, to complain. 'Don't overreact', advised Powell, 'that magazine has been on the new-stands for a week. Other magazines are full of maps showing other battle plans. They're all just speculating.' Schwarzkopf was relieved to see that Powell was right and there was no subsequent change in Iraqi dispositions.[7]

The Coalition meanwhile had considerable intelligence about Iraqi dispositions, and as the air and Psyops campaigns progressed, so the large numbers of defectors provided further evidence of both Iraqi planning and the condition of their troops. Quite apart from the leaflets urging defection, the accuracy of the bombing and the ability to destroy hardware provided a powerful psychological weapon. Over Christmas, Schwarzkopf and his staff watched video tapes of the Ken Burns film *The Civil War*, which served to bring home to them the magnitude of the task ahead. It renewed Schwarzkopf's determination to do everything possible to minimize the loss of life.[8] He called his subsequent deception plan the 'Hail Mary Play', and it involved going around the Iraqi defences rather than through them.

The US Marine Corps would demonstrate as if to make a landing along the coast, while XVIII and VII Corps would redeploy from near the coast to positions well to the west, from which they would sweep around the back of the Iraqi defence line. Positions would be held for as long as possible before making this manœuvre, and dummy headquarters would remain to create electronic signatures in the old locations. At the same time, in press briefings, massed assaults and breaching tactics were described, and a subtle spin imparted that implied a direct attack combined with amphibious operations in support. Meanwhile, Psyops materials being used included leaflets showing US Marines surfing onto the beaches with helicopter and naval support, highlighting the real capabilities without comprising genuine intentions.

On 16 January 1991 all Coalition ground forces were arrayed east of a line running through Hafr Al Batin to the Saudi–Iraqi–Kuwaiti tri-border junction, squeezing them closely together. The first task of the air campaign was to neutralize the Iraqi air force and deny it the ability to see what the Coalition was doing. Covered by massive air and artillery attacks, USMC formations swapped places with Saudi and Kuwaiti forces in the east and VII Corps moved to the end of the Saddam Line, while XVIII Corps made a massive move westwards, covering an average distance of 360 miles. To ensure it was not detected, it was held south of the Tapline Road and, since it was feared that nomadic Bedouin might compromise the security of this manœuvre, Saudi units were sent in first to clear as many from the area as possible.

The VII Corps moved an average of 140 miles to its left, leaving a conspicuous gap between it and XVIII Corps in order to suggest that this was the end of the line. Many elements were held back until shortly before the ground invasion opened. An ammunition dump covering forty square miles was created, but the setting up of two other large logistic bases was deferred until Iraqi air recce had been neutralized. The XVIII Corps left a hundred-man deception cell in eastern Saudi Arabia to use inflatable decoys and radio measures.[9] The large armoured VII Corps waited until 16 February to commence its move and also left behind a deception cell, which created a complete decoy military base south of Wadi Al Batin with mock missiles, fuel dumps, radio traffic, Hawk missile radar signals, and vehicles using multi-spectral decoys to make it harder for the Iraqis to see that all of VII Corps was moving westwards.[10]

Between 17 January and 17 February the Coalition moved over 100,000 men and 1,200 tanks as well as thousands of other vehicles, an enormous logistical effort. Schwarzkopf wanted sufficient food, ammunition, spares and other supplies to last this force for sixty days, which required the construction of three vast depots and a torrent of traffic (one truck passing any given point along the two-lane Tapline Road every fifteen seconds). This could only be achieved with total control of the air, although the air campaign's focus remained Kuwait and the area to the immediate west (which further sustained the deception plan). Similarly, skirmishing was maintained along the Kuwaiti border and just west of Kuwait counter-reconnaissance raids were carried out by elements of the US 1st Cavalry and 1st Infantry Divisions after 9 February. Further deceptive relocations occurred after this point when 1st USMC Division, which had previously been deployed opposite the A1 Wafra oilfields, moved rapidly to a position opposite the bend in the Kuwaiti border and 2nd USMC Division, which had been stationed east of the 1st, moved to new positions further west.

The 1st (UK) Armoured Division was regarded as a key signature formation, whose position would give the Iraqis a clear indication of the direction of the main effort. Seventh Armoured Brigade and later 4th Armoured Brigade were originally under command of the Marine Expeditionary Force (MEF) USMC, and was also to move west along the Tapline Road. The Iraqi Army had fourteen electronic warfare battalions, of which nine were believed deployed in the Kuwait theatre. Therefore the Fleet Electronic Warfare Support Group (FEWSG, the Royal Navy's communications security monitoring team), recorded radio transmissions from Sherpa vans, and when the division moved westwards, a group call RHINO Force comprising half a dozen matelots and transmitter equipment borrowed from the British Forces Broadcasting Service (BFBS), remained in place to rebroadcast the exercise traffic that had been generated. At the same time the resubordination of 1st (UK) Armoured Division from the Marines to VII Corps on 26 January was not made public, and a television report showed 26th Field Regiment, Royal Artillery, training on a range by the sea, without stating that the regiment had only just arrived and would be moving to join the rest of the division almost immediately.

The amphibious deception had begun as early as August 1990, when 2,500 Marines sailed for the Mediterranean on the amphibious assault ship USS *Inchon*. Meanwhile, elements from Diego Garcia, Guam and the Atlantic were directed towards the Persian Gulf. In mid-August the Pentagon announced the dispatch of a 15,000-man Marine Expeditionary Brigade (MEB) aboard some thirteen ships, and this force henceforth received prominent press coverage. By mid-January the total Marine deployment was around 17,000 men from two MEBs and two Amphibious Task Groups.[11] On 17 January the Coalition announced publicly that air attacks would begin to soften up the coast defences, and on the 25th that the current amphibious exercises were the largest 'since Korea'. On the 28th an attack on Iraqi warships in Kuwaiti waters was made, ostensibly to open an approach for a Marine assault and US Navy sea–air–land teams (SEALs) to conduct mine-clearing operations, while battleships hammered coastal defences. At the same time rehearsals continued: in the last ten days in January 8,000 Marines exercised in Oman. On 1 February *Newsweek* magazine carried a feature article on the planned amphibious assault. By 22 February up to 80,000 Iraqis were reported as defending the beaches. On the 25th Marine helicopters flew a series of missions along the coast while 13th Marine Expeditionary Brigade feigned an attack. On the first day of the ground offensive Radio Free Kuwait claimed that Marines had landed on Faylakah Island, a deception that was maintained for some time afterwards.[12]

Once the ground campaign opened, the other deception operations also continued. The two Marine divisions attacked at 0400 hours on 24 February in the eastern end of the sector, where Coalition planners wanted the Iraqis to think all the assaults would take place before the assault commenced in the west, and this successfully breached two lines of defence. All the Coalition forces demonstrated vigorously in this sector; most notably, 1st Cavalry Division launched a mock attack just west of Wadi A1 Batin to fix the Iraqis in that area. Meanwhile, XVIII Corps launched itself deep into the desert to establish forward staging areas with the French 6th Light Armoured Division seizing Salmon airstrip, and US 101st Airborne Division blocking Highway 8. So successful were all these operations that VII Corps, originally not due to cross its start lines for twenty-four hours, was ordered forward by Schwarzkopf during the afternoon of 24 February. By this stage there was practically nothing the Iraqi High Command could do to reorientate itself to the real threat. Crammed as they were into a 200-mile wedge along the southern border and eastern coastline, they offered scant resistance to the Coalition forces making the main effort out to the west.

It would be a mistake to suppose that the ubiquity of electronic media made large-scale operational deception of this sort difficult for the Coalition. The Iraqis sought to exploit the presence of CNN in Baghdad for their own purposes, but this proved a double-edged sword. The emphasis on capabilities helped concentrate the minds of Iraqi planners on those aspects which Coalition planners wanted them to focus on, such as the amphibious threat, without mentioning intentions. The only difference between this and previous deception schemes is one of emphasis: Schwarzkopf fed the news-hungry media plenty of information that was not false, merely skewed to suit his purposes. In so doing, Coalition planners showed real understanding and skilful appreciation of the nature of modern mass media. By the same token, every utterance made by President Bush and other Western political leaders emphasized the limitations of their mandate from the UN as being the liberation of Kuwait, thus making the militarily implausible option of attacking the Iraqis where they were strongest appear plausible.

The abundance of information regarding orders of battle in modern professional armies meant that while the location of the assault was successfully hidden, other aspects were not so effectively obscured. In fact, there was almost a catastrophic security breach when a laptop computer containing details of the plan was stolen from a parked car belonging to an assistant of a very senior British officer in London. Fortunately, the computer was handed in anonymously to the Ministry of Defence three weeks later, and while there is no evidence that any of the details were leaked to the Iraqis, that can only be attributed to good fortune.[13]

COUNTERING AND TEACHING DECEPTION

Self-delusion on the part of the target is undoubtedly a major factor in successful deception. Complacency and over-confidence, in particular, can make a force vulnerable. During the Gulf War, for example, the Coalition was surprised by the Iraqi raid on Khafji, perhaps because of the Iraqis' otherwise inept conduct. The modern belief in the infallibility of SIGINT is another example of over-confidence, since this is the most easily fabricated form of intelligence. Besides, the signals world of the twenty-first century is very different from that of the mid-twentieth. The electronic environment is now so densely packed with transmissions that sophisticated judgement is needed to choose which signals are actually worth recording, even before any code-breaking takes place. By the same token, increasing sophistication of military equipment makes fire-control radar, for example, difficult to identify owing to the ability to shift wavelength and waveform, a characteristic developed to guard against electronic counter-measures. In signals intelligence, meaning is much more elusive than imagery.[14]

The standard safeguard against deception is to check how well the new information or interpretation fits with the existing broad picture of reality. Intelligence, however, looks at very small and exceptionally important pieces of reality using narrow, specialist sources and methods, and this specialization makes the task of checking the image with the broader world all the more difficult. Technology can provide glimpses of otherwise denied areas but may not produce useful information without some broader understanding of the enemy's intentions. Moreover, any nation that knows it is being observed in this way will naturally seek to control and manipulate the information available to the enemy's technology.[15] Simply possessing high-technology information- gathering systems does not guarantee high-quality information.

Counter-measures to deception are necessary and should be treated as seriously by modern commanders as their own deception schemes. Commanders should be under no doubt that they will be deception targets themselves, and the greater their own predilection for trickery, probably the greater their own awareness of that likelihood. Good intelligence will lead them some way towards uncovering a deception, and intelligence staff need to be able to distinguish deceptive threats from genuine ones. They must therefore be sceptical about every item of information that they receive, and must never accept anything purely at face value. And intelligence officers must resist the temptation to produce conclusions for which they have only hunches (and in the case of 'known' doctrine on the part of forces trained by the former USSR, it would be very foolish simply to ascribe to them the behaviour patterns described in intelligence manuals).

In this respect, the current American system of Intelligence Preparation of the Battlefield (IPB) is ripe for enemy deception planners. By use of overlays, it attempts to simplify the intelligence estimate process and, especially in defence, it relies heavily on a stylized portrayal of the Soviet pattern of operational behaviour.[16] Its use of templates is simply far too prescriptive, particularly where it templates doctrine (a practice with inherent dangers that are likely to be exacerbated in the absence of hard intelligence) and assumes little or no operational flexibility or imagination on the opponent's part, whereas identification and use against an enemy of his 'specified algorithms of decision' are implicit in the very concept of *maskirovka*.[17] In other words, anyone with an understanding of deception, including those schooled in *maskirovka*, will seek to exploit the limitations of IPB by feeding what its practitioners might expect to see and hear. During the Great Patriotic War the Soviets repeatedly chose as an axis terrain dismissed by the Germans as unsuitable for a major offensive, much as the Germans themselves did in choosing the Ardennes in May 1940 and, perhaps more significantly, in December 1944. In Soviet eyes the best terrain was that in which there were no anti-tank weapons.[18]

A skilful attacker will therefore tailor his deception to confirm the view the defender already has of him before doing something completely different, much as the Egyptians did in 1973. It is therefore imperative for both commanders and staffs to keep an open mind, particularly to be aware as far as possible of their own preconceptions, so that these are not exploited. The only way to learn how to counter deception is to teach to it and to exercise it, through practice and at all levels. This means introducing realistic deceptive elements into both sides during training exercises, rather than merely discussing them as an afterthought to an exercise instruction.

If the opposition force in an exercise is operating to a strict schedule provided by an umpire, it will make for dull and unrealistic training. Furthermore, as Basil Liddell Hart wrote in his *Strategy of Indirect Approach*:

> the training of armies is primarily devoted to developing efficiency in the detailed execution of tactics. The concentration on tactical techniques tends to obscure the psychological elements. It fosters a cult of soundness rather than surprise. It breeds commanders who are so intent not to do anything wrong, according to 'the book' that they forget the necessity of making the enemy do something wrong.

This remains true to this day. Tactics are taught in a way that seeks to exploit the best ground, but if the enemy knows what the best ground is he is likely to plan accordingly. Choosing second-best terrain, by contrast, creates an opportunity for a deception that may yield enormous results: surprise and consequent success at quite possibly much less cost.

Training in deception involves teaching commanders to make the enemy make mistakes, a task that requires imagination. Some will naturally show a greater talent for it than others, but this talent can only be encouraged by allowing as much lateral thinking as possible into training schemes, which all too often are structured and formulaic. Wavell ran exercises based on the legend of the Golden Fleece, on *King Solomon's Mines* and on other ideas gleaned from books and films.[19] His aim was to make training unusual and, as a result, more interesting and more effective. Most Western soldiers, however, tend to consider personal camouflage as the limit of their deceptive responsibilities, and deception is usually passed over on exercises with the excuse that resources are scarce. (Surely this is itself an excellent argument in favour of practising deception at every level and at every opportunity.) According to modern British Army doctrine, effective deception requires the commitment of significant resources to convince an enemy. One can only assume that its authors have never studied the art of deception.

An example of this problem is the use of smoke. With their experience of warfare on the steppes, it is hardly surprising that the Soviets appreciated the value of smoke. As well as being used for concealment, smoke served to draw fire from German positions and thus assist in reconnaissance. Soviet thinking on the use of smoke changed very little in the post-war years. The Yom Kippur War re-emphasized its value and it was always considered a low-technology counter-measure to Western high-technology weapons. Whatever other deficiencies may have been noted in the practice of *maskirovka* on exercises, use of smoke was not among them. Soviet tanks were all designed with the ability to inject a chemical mixture into their exhausts and create their own smoke continuously. The best most Western tanks can manage is a few smoke rounds from a discharger. In Soviet and Warsaw Pact exercises smoke was always used in abundance and it was found that it could greatly reduce casualties. Smoke capable of blinding thermal-imaging devices can be produced and screens of enormous dimensions

Maryland. They predict the ability to create holographic forces, a concept inspired by the original *Star Wars* film, when a message was delivered by a holographic image.[21] *Star Wars* also inspired the Imperial trooper, and continuing developments in miniaturization will, it is predicted, produce similar soldiers equipped with helmets including head-up displays, global positioning, video and other sensor links direct to headquarters, and wearing protective suits with lightweight body armour, laser-guided weapons and other such gadgetry. But while modern warfare is undoubtedly seeing an enormous increase in the quantity of information, there is also a marked reduction in the time available to intelligence staffs to analyse it and disseminate the resulting intelligence. Modern deception measures should take advantage of the target's vulnerability to paralysis through data saturation. A Soviet writer noted the necessity for

> dummy objectives to possess the physical properties of the equipment being simulated . . . not only having the appropriate form, but also being capable of reflecting any light, heat and electromagnetic energy which falls on them, and also creating heat emissions, a magnetic field around themselves etc. Otherwise modern means of reconnaissance will differentiate with relative ease between the true and false targets.[22]

Although the use of special paints, radar reflectors and other measures to create deceptive displays of vehicles or other equipment has not been confined exclusively to the former USSR and Warsaw Pact, these nations retained the willingness to use such tactics on a colossal scale that NATO perhaps never fully understood. It is not unreasonable to assume that wherever Russian equipment is sold and advice given these techniques will be among them.

In any case, no matter how radical a technological innovation may be, a counter-technology will soon be developed. Infra-red sensors designed to detect heat can be countered by something as simple as treated hessian material, used in conjunction with camouflage netting. Similarly, sensors cannot differentiate between a genuine hot engine and a simulation made from a can filled with sand and petrol. Developments in 'stealth' technology are taking place to make equipment invisible to radar. Conversely, one vehicle can tow a string of reflectors behind it to simulate many others. From the deceiver's point of view, more spectra have to be covered than before, but the means are there, as demonstrated by recent experience.

Under some circumstances the latest technology can remain vulnerable to surprisingly unsophisticated deception strategies. During the Kosovo conflict very simple decoys and dummy sites led the most powerful air forces in the world to overestimate grossly the effectiveness of their campaign against Serb armour on the ground. And since NATO aircraft delivered most of their attacks from over 15,000 feet in order to avoid anti-aircraft fire, it is hardly surprising that in spite of repeated claims about accuracy they sometimes hit refugee columns rather than military convoys. A British ordnance officer reported finding the remains of only thirteen Serb tanks destroyed from the air – the same number as the Serbs admitted to during the conflict and eighty- three fewer than NATO claimed at the time. NATO pilots were simply fooled into attacking hundreds of other locations.[23] An internal Pentagon report obtained by *Newsweek* revealed that only 58 strikes were accurate, compared with 744 'confirmed' by NATO at the end of the campaign. Far from destroying 'around 120 tanks', 'around 220 armoured personnel carriers (APCs)' and 'up to 450 artillery and mortar pieces', in seventy-eight days of bombing, the true figures were 14 tanks, 18 APCs and 20 guns and mortars. Another unwelcome discovery was just how easily high-altitude surveillance systems had been tricked. Yet the Serbs did not possess state of-the-art camouflage and deception equipment. Although their battle positions were locatable, their hides seldom were, and many of the decoys were just knocked together from local materials. 'Tanks', for example, were made from black logs on old lorry wheels. (Other characteristics can be simulated just as easily: a petrol can will make the 'tank' explode with an appropriate whump and chicken wire can give it a measure of protection against certain anti-tank missiles.) One important bridge was 'destroyed' many times over, protected by a fake built 300 metres away from polyethylene sheeting.[24] Another bridge at Kosovo Polje was constructed from local timber and roofing felt, which has the same thermal signature as tarmac and which provided a perfectly convincing roadway.

American military professionals have accepted their deficiency in deceptive techniques for some years. This failing is perhaps a result of urbanization, and possibly a product of having fought the last five major wars and other smaller campaigns with overwhelming air superiority.[25] While that superiority is in itself extremely desirable, it is not everything. Satellites also have severe limitations, depending on the height of orbit. In a geosynchronous orbit at 22, 300 miles above the earth, a fairly large area of the earth can be seen, but very little detail can be distinguished. At an altitude of 100 miles, on the other hand, a satellite can see a duck on a pond, but would circle the earth in little more than a hour and cover a very limited area. Thousands of such satellites would be needed to guarantee meaningful coverage, involving expense that no country can afford; in any case, looking for tanks and missiles may be fruitless since these can always be kept indoors or under cover.[26] High technology is not a panacea nor an end in itself; more traditional human intelligence sources may yet have a significant role to play in the future.

Doctrinally, deception now forms one of the pillars in the concept of command and control warfare (C2W), which aims to attack an enemy's command and control functions (and to protect one's own) through the combined and co-ordinated use of operational security, physical destruction, deception, electronic warfare and Psyops. These techniques are part of what is now regarded as 'information age warfare', but it is important to distinguish between this and 'information warfare'. C2W differs from information warfare in that the former concerns only the enemy's military capabilities while the latter focuses on any information function, including the civilian infrastructure (which also formed an important part of the NATO attack on Serbia).[27] Information-age warfare incorporates new technologies into the actual business of directing and implementing military operations, while information warfare views information in itself as a 'separate realm, potent weapon and lucrative target'.[28]

Although it foresees the use of high-technology techniques such as 'logic' bombs to attack the information systems on which everybody increasingly relies, contemporary warfare also includes conventional attacks on information functions, such as telephone switching facilities. The array of potential targets is vast, and the greater a nation's reliance on high technology, the greater its vulnerability to such attacks. Consequently, the use of deception may have an important part to play in protecting these vulnerable functions, by suggesting, for example, that an attack has been more successful than is really the case.

Direct information warfare affects information without relying on the opponent's ability to perceive or interpret that information. Attacks might take many forms, but those of a deceptive nature could involve planting false information in an enemy's databases (creating a false order of battle, for example). Nevertheless, although such schemes involve new methods, the time-honoured principles remain entirely valid. If a fighter wing appears in this new order of battle the enemy will seek corroboration, and false runways and decoy aircraft (with all the trimmings) are as necessary as ever. Alternatively, if dummy bridges prevent the destruction of a real bridge, an information attack might achieve the desired effect by deceiving the enemy into thinking that a targeted bridge has been destroyed, causing him to reroute transport and divert engineer resources for its repair. This remains a largely hypothetical and speculative area, but it is only by carefully considering all such means that defences will be developed.

Postscript

It is not unreasonable to say that the more expensive and extensive wars become, the more valuable successful deceptions are likely to become.[29] Wars continue around the globe, but many 'small-scale' conflicts excite little or no media interest in the West. When the West does become involved, it increasingly relies on its huge technological advantage. This is to its benefit only so long as it remembers that wars are fought not by machines but by men; and the best soldiers have a seasoning of devilry.[30]

However, it is worth sounding a final note of caution. The exact success of any given deception operation is often very difficult to measure. There are widely varying estimates of the success of the 'going-map' ruse (in which a map that marked poor terrain as 'good going' was planted on the Germans before the Battle of Alam Haifa). And Klaus-Jürgen Müller argues that many writers have overestimated the effectiveness of Allied deception operations during the Second World War, including BARCLAY and FORTITUDE NORTH.[31] By the same token, a sense of proportion is needed: deception is probably less important than good intelligence, and no war was ever won by either, but only by hard fighting.

As Sherman said on a number of occasions, 'War is hell and you cannot refine it.' Nobody with any experience of war would seek to disagree, yet, as Dudley Clarke wrote in the foreword to his draft memoirs (sadly never completed or published), 'the secret war was waged rather to conserve than to destroy; the stakes were the lives of the frontline troops, and the organization which fought it was able to count its gains from the number of casualties it could avert.'[32] Deception remains a powerful tool for reducing the bloodshed inherent in war, as well as having an enormous influence on its outcome. The art of deception in war is far from dead.

NOTES

INTRODUCTION

1. *Design for Military Operations: The British Military Doctrine,* pp. 4–35.
2. D. Glantz, *Soviet Military Deception in the Second World War,* p.2.
3. A. P. Wavell, *The Good Soldier,* pp. 157–8.
4. Polybius, *The Rise of the Roman Empire,* III, 81, p.248.
5. D. Mure, *Practise to Deceive,* p.14.
6. R. Beaumont, *Maskirovka: Soviet Camouflage, Concealment and Deception,* p.21.
7. D. Glantz, p.3.
8. C. Cruickshank, *Deception in World War II,* p.123.
9. M. Hastings, *Overlord,* p.75.

CHAPTER 1. A HISTORY OF BLUFF IN WARFARE

1. Judges, vii, 15–23. See also Isaiah, ix, 4.
2. Sun Tzu, *The Art of War,* p.66.
3. Polybius, *The Rise of the Roman Empire,* III, 46, pp.218–19.
4. Julius Caesar, *The Conquest of Gaul,* VII, 35, p.172.

5. M. I. Handel (ed.), *Strategic and Operational Deception in the Second World War,* p.2.
6. A. P. Wavell, *The Good Soldier,* p.36.
7. Sir C. Oman, *A History of the Art of War in the Middle Ages,* pp.172, 201–5.
8. R. Holmes, *War Walks* 2, p.37.
9. H. Delbrück, *History of the Art of War, vol. 3, The Middle Ages,* p.159.
10. Sir C. Oman, p.162.
11. J. Chambers, *The Devil's Horsemen,* pp.79–80.
12. D. Morgan, *The Mongols,* p.86.
13. R. E. Dupuy and T. N. Dupuy, *The Collins Encyclopaedia of Military History,* pp. 372–3.
14. P. Thorau, The Lion of Egypt*: Sultan Baybars and the Near East in the Thirteenth Century,* pp.204–5.
15. G. A. Williams, *Owain Glyndŵr,* pp.27–8.
16. J. D. G. Davies, *Owen Glyn Dŵr,* pp.45–50.
17. G. Hartcup, *Camouflage,* p.11.
18. N. Machiavelli, *The Art of War,* p.xxv.
19. H. Delbrück, vol. IV, *The Dawn of Modern Warfare,* p.156.
20. D. Chandler, *Marlborough as Military Commander,* pp.126–30.
21. M. Dewar, *The Art of Deception in Warfare,* p.31.
22. D. McKay, *Prince Eugene of Savoy,* p.59; N. Henderson, *Prince Eugen of Savoy,* p.58.
23. D. Chandler, *Marlborough,* pp.158–61.
24. Ibid., pp.287–91.
25. J. Luvaas (ed.), *Frederick the Great on the Art of War,* p.324.

26. C. Duffy, *Frederick the Great: A Military Life*, p.327.

27. D. Chandler, *Atlas of Military Strategy: The Art, Theory and Practice of War, 1618–1878*, p.73.

28. C. Duffy, pp.148–53.

29. A. V. Sellwood, *The Saturday Night Soldiers, p.143.*

30. *London Gazette Extraordinary (27 March 1797); Annual Register* (1797); *quoted in* E. H. Stuart-Jones, *The Last Invasion of Britain*, p.114.

31. Ibid., p. 115.

32. Ibid., p.111.

33. Ibid., p.116–19.

34. D. Chandler, *The Campaigns of Napoleon*, p.146. For a detailed exposition of what follows, this is the definitive reference.

35. D. Chandler, *Atlas*, p.98.

36. D. Chandler, *Campaigns of Napoleon*, pp.78–80.

37. M. van Creveld, *Supplying War*, pp.40–2.

38. D. Chandler, *Atlas*, p.100–1.

39. A. P. Wavell, p.38.

40. N. Dixon, *On the Psychology of Military Incompetence*, p.326.

41. D. R. Morris, *The Washing of the Spears*, p.53.

42. Ibid., p.63.

43. G. C. Ward et al., *The Civil War*, pp.76–80. An excellent overview taken from the film of the same name, and lavishly illustrated.

44. Ibid., p.90.

45. P. Katcher, Great *Gambles of the Civil War*, p.195.

46. G. C. Ward et al., p.110.

47. P. Katcher, p.202.

48. W. J. Miller, 'No American Sevastopol', *America's Civil War,* vol. 13, 2, p.34.

49. C. V. Woodward, *Mary Chesnut's Civil War,* p.401.

50. S. Foote, *The Civil War: A Narrative,* vol. 1, pp.381–4. A wonderful book, beautifully written.

51. G. C. Ward et al., p.124.

52. Ibid., p.340.

53. J. A. Wyeth, *Life of General Nathan Bedford Forrest,* pp.108–9.

54. Ibid., p.218.

55. S. Foote, vol. 2, p.185.

56. J. A. Wyeth, pp.492–3.

57. T. Pakenham, *The Boer War,* p.398.

58. R. S. S. Baden-Powell, *Lessons from the 'Varsity of Life,* p.203.

59. R. S. S. Baden-Powell, evidence, *Royal Commission on the War in South Africa,* vol. 3, p.424.

60. R. S. S. Baden-Powell, *Lessons from the 'Varsity of Life,* p.202.

61. D. Grinnell-Milne, *Mafeking,* pp.50–2.

62. R. S. S. Baden-Powell, *Lessons from the 'Varsity of Life,* p.211.

63. *B. Gardner, Mafeking: A Victorian Legend, p.65.*

64. R. S. S. Baden-Powell, *Lessons from the 'Varsity of Life,* p.206.

65. D. Grinnell-Milne, p.58.

66. Ibid., p.94.

67. A. Home, *The Price of Glory,* pp.37–40.

Chapter 2. The Information Battle

1. See A. D. Coox, 'Flawed Perception and its Effect upon Operational Thinking: The Case of the Japanese Army, 1937–41', in M. I. Handel (ed.), *Intelligence and Military Operations.*

2. K. von Clausewitz, *On War,* p.162.

3. See Chapter II, 'Seurat and his Friends', in J. Rewald, *Post Impressionism: From Van Gogh to Gauguin,* London, Secker and Warburg, 1978.

4. R. K. Betts, 'Analysis, War and Decision: Why Intelligence Failures Are Inevitable', *World Politics* (31 October 1978), pp.69–72.

5. Sir J. Ardagh, evidence, *RCSAW* II, Q5126, quoted in Pakenham, p.76.

6. D. Chandler, *Marlborough as Military Commander,* p.324.

7. J. Luvaas (ed.), *Frederick the Great on the Art of War,* p.122.

8. J. G. Zimmerman, *Fragmente über Friedrich den Grossen,* vol. 1, Leipzig, 1790, p.288.

9. C. Duffy, *Frederick the Great: A Military Life,* pp.326–7.

10. D. Chandler, *The Campaigns of Napoleon,* pp.146–7.

11. S. Foote, *The Civil War: A Narrative,* vol. 3, p.144.

12. Frontinus, *Stratagems,* p.21.

13. S. Foote, vol. 2, p.689.

14. C. Roetter, *Psychological Warfare,* pp.136–7.

15. S. Foote, vol. 1, p.570.

16. J. Haswell, *The Tangled Web: The Art of Tactical and Strategic Deception,* pp.89–90.
17. M. de Arcangelis, *Electronic Warfare,* pp.11–12.
18. B. W. Tuchman, *The Guns of August: August 1914,* pp.327, 344.
19. 'A History of Electronic Warfare', Royal School of Signals pamphlet.
20. For a full description of these German operations, see H. O. Behrendt, *Rommel's Intelligence in the Desert,* London, William Kimber, 1980. The childishness of British radio procedure is well illustrated in K. Douglas, *Alamein to Zem Zem,* Harmondsworth, Penguin, 1969.
21. G. Barkas, *The Camouflage Story,* p.43.
22. A. J. Brookes, *Photo Reconnaissance,* pp.9–10.
23. B. W. Tuchman, pp.270–1.
24. A. J. Brookes, p.13.
25. Ibid., pp.16–20.
26. C. Babington-Smith, *Air Spy,* p.113.
27. J. Haswell, p.111.
28. A. J. Brookes, pp.83, 89–90.
29. D. Chandler, pp.146–8.
30. G. Barkas, p.50.
31. Polybius, *The Rise of the Roman Empire,* III, 78, p.245.
32. B. Pitt, *The Crucible of War,* pp.81, 85, 90.
33. 'The Principles and Practice of Camouflage', TM 623.77(41) Camouflage/1, p.7.
34. T. Newark, Q. Newark and J. F. Borsarello, *Brassey's Book of Camouflage,* p.11.

35. I. T. Schick (ed.), *Battledress: The Uniforms of the World's Great Armies, 1700 to the Present,* pp.145–6.

36. T. Newark et al., pp.12–14.

37. I. T. Schick, pp. 189, 210.

38. B. W. Tuchman, pp.55.

39. I. T. Schick, pp.219–20.

40. T. Newark, et al., pp.15–16.

41. G. Hartcup, *Camouflage: A History of Concealment and Deception in War,* pp.77–86.

Chapter 3. The Principles of Deception

1. PRO AIR 20/2497, 05.02.1944.

2. D. Mure, *Master of Deception,* p.81–2.

3. PRO CAB 154/2, 'A' Force War Diary, vol. 2, p.16.

4. J. Haswell, *The Tangled Web: The Art of Tactical and Strategic Deception,* p.23.

5. PRO CAB 154/2, p.19.

6. R. Hesketh, *Fortitude,* pp.360–1.

7. P. Beesley, *Room 40: British Naval Intelligence, 1914–18,* p.184.

8. J. Haswell, pp.85–8.

9. R. Hesketh, pp.360–1.

10. M. Howard, *British Intelligence in the Second World War, vol. 5, Strategic Deception, p.105.*

11. D. Mure, *Practise to Deceive, p.14.*

12. R. Hesketh, pp.360–1.

13. PRO CAB 154/2, p.22.

14. A. M. Codevilla, 'Space, Intelligence and Deception', in B. D. Dailey and P. J. Parker (eds.), *Soviet Strategic Deception*, p.484.

15. D. Glantz, *Soviet Military Deception in the Second World War*, pp.568–9.

16. M. Young and R. Stamp, *Trojan Horses: Deception Operations in the Second World War*, p.71.

17. B. Whaley, *Stratagem*, p.229. Whaley also discusses the relationship between security and deception with different emphases, pp.225–6.

18. W. Shirer, *The Rise and Fall of the Third Reich*, pp.671–2.

19. R. Hesketh, p.355.

20. NA C. H. Bennett, 'German Appreciation of Operation STARKEY', COSSAC/41 DX/INT, 01.09.1943, R331.

21. J. Haswell, p.24.

22. C. Cruickshank, *Deception in World War II*, p.i.

23. R. Hesketh, p.356.

24. D. Mure, *Practise to Deceive*, p.49.

25. D. C. Daniel and K. L. Herbig, 'Propositions on Military Deception', in *Strategic Military Deception*, pp.18–19.

26. D. Mure, *Master of Deception, p.14.*

27. J. C. Masterman, *The Double Cross System*, p.110.

28. D. C. Daniel and K. L. Herbig, p.20.

29. R. Lewin, *Ultra Goes to War*, pp.237, 316.

CHAPTER 4. THE METHODS OF DECEPTION

1. D. C. Daniel and K. L. Herbig, pp.5–7.

2. A. P. Wavell, *The Good Soldier*, p.157.

3. J. Haswell, *The Tangled Web: The Art of Tactical and Strategic Deception, pp.30–2.*
4. S. Foote, *The Civil War: A Narrative,* vol. 2, p.332.
5. Frontinus, *Stratagems,* pp.75–7.
6. NA APO #655, RG319, G-3, 11.07.1944.
7. S. Foote, vol. 2, pp.685–8.
8. M. Young and R. Stamp, *Trojan Horses: Deception Operations in the Second World War,* pp.86–8.
9. PRO WO 201/2649.
10. J. Luvaas, *Frederick the Great on the Art of War, pp.122–3.*
11. D. Mure, *Master of Deception, p.27.*
12. R. Hesketh, *Fortitude,* pp.351–3.
13. Ibid., pp.15, 65–71.
14. D. Mure, *Practise to Deceive,* p.49–50.
15. D. Mure, *Master of Deception,* pp.76–8.
16. D. Mure, *Practise to Deceive,* p.14.
17. R. Hesketh, p.38.
18. Ibid., p.358.
19. PRO FO 898/398, 29.06.1943.
20. C. Cruickshank, *Deception in World War II,* p.56.
21. A. Cave Brown, *Bodyguard of Lies,* p.118.
22. PRO WO 201/2023.
23. G. Barkas, *The Camouflage Story,* pp.198–200, 204–5.
24. M. Young and R. Stamp, p.70–1.
25. P. Delaforce, *Monty's Marauders,* pp.44–6.
26. D. Mure, *Master of Deception,* pp.126, 128.
27. G. Barkas, pp.210–11.
28. Ibid. pp.200–3.
29. D. Mure, *Master of Deception,* p.135.

30. PRO WO 201/2024.

31. PRO WO 201/2023.

32. M. Dewar, *The Art of Deception in Warfare,* p.10.

33. *Sunday Times Insight Team, The Yom Kippur War,* p.61.

34. Ibid., p.69.

35. J. Amos, 'Deception and the 1973 Middle East War', in D. C. Daniel and K. L. Herbig (eds.), *Strategic Military Deception,* pp.322–4.

36. Ibid., pp.324–6: R. K. Betts, *Surprise Attack,* p.72.

37. J. Amos in D. C. Daniel and K. L. Herbig, pp.327–8.

38. J. Lucas, *Kommando: German Special Forces of World War Two,* p.45.

39. I. C. B. Dear, ed., *The Oxford Companion to the Second World War,* p.155.

40. M. R. D. Foot, *Resistance,* p.29.

41. P. Fleming, *Operation Sealion,* pp.63–4.

42. J. Lucas, pp.48–50.

43. Ibid., pp.129–30.

44. L. Kessler, *Kommando: Hitler's Special Forces in the Second World War, pp.111–16.*

45. J. Lucas, pp.131–2.

46. L. Kessler, p.129.

47. A. P. Wavell, *The Palestine Campaigns,* pp.100–3.

48. There are a variety of accounts of Meinertzhagen's 'haversack ruse': see M. Dewar, *The Art of Deception in Warfare,* pp.39–40, and M. I. Handel (ed.), *Strategic and Operational Deception in the Second World War,* p.8. See also C. Falls, *Military Operations Egypt and Palestine,* part 1, London, HMSO, pp.30–31; Sir G.

Aston, *Secret Service,* New York, Cosmopolitan, 1930, pp.201–16.

49. A. P. Wavell, p.106–7.

50. M. I. Handel, pp.10–11.

51. M. Howard, *British Intelligence in the Second World War: Strategic Deception,* p.89.

52. E. Montagu, *The Man Who Never Was,* p.13.

53. Ibid., p.54.

54. H. Greiner and P. Schramm, *Kriegstagebuch des Oberkommando der Wehrmacht,* vol. 3, p.1429.

55. M. Howard, p.92.

56. J. Pimlott, *Vietnam: The Decisive Battles,* pp.90–1.

57. J. Morrocco et al., *The Vietnam Experience: Thunder in the Air,* pp.144–6.

58. J. Pimlott, p.92.

59. C. Cruickshank, p.88.

60. Livy, *The History of Rome,* xxii, 14.

61. L. Cottrell, *Enemy of Rome,* p.114.

62. J. Peddie, *Hannibal's War,* pp.84–5.

Chapter 5. Tactical and Operational Deception

1. F. Wilson, *Regiments at a Glance,* p.53.

2. PRO AIR 23/5526, 17.03.1943.

3. A. Gilbert, *Sniper: One-on-One,* p.71.

4. H. St G. Saunders, *The Red Beret,* p.257.

5. G. Hartcup, *Camouflage,* p.27.

6. C. Wilmot, *The Struggle for Europe,* pp.406–8.

7. P. Delaforce, *The Fighting Wessex Wyverns,* p. 110.

8. G. Taylor, *Infantry Colonel,* p.66.

9. Ibid., pp.67–72.
10. H. St G. Saunders, pp.283–4.
11. R. Beaumont, *Maskirovka: Soviet Concealment, Camouflage and Deception*, p.9.
12. J. Lucas, *War on the Eastern Front, 1941–45: The German Soldier in Russia*, pp.53–5, 96, 181.
13. F. W. von Mellenthin, *Panzer Battles: A Study in the Employment of Armor in the Second World War*, p.294.
14. S. H. Newton (ed.), *German Battle Tactics on the Russian Front, 1941–5*, p.72.
15. Ibid., p.77.
16. TM 623.77(41). Camouflage/1, pp.12–13.
17. J. Maskelyne, *Magic: Top Secret*, p.47.
18. S. Reit, *Masquerade: The Amazing Camouflage Deceptions of World War II*, p.179.
19. Ibid., p.187.
20. G. Hartcup, pp.111–12.
21. A. J. Barker, *Fortune Favours the Brave: The Battle of the Hook, Korea, 1953, pp.77–85.*
22. A. Moorehead, *Gallipoli,* p.263.
23. M. Hickey, *Gallipoli,* pp.322, 327–9.
24. A. Moorehead, p.287.
25. Ibid., p.288.
26. M. Hickey, pp.329–34.
27. J. Haswell, *The Tangled Web: The Art of Tactical and Strategic Deception, pp.75–6.*
28. *The Times,* 14 May 1974.
29. D. Mure, *Master of Deception*, p.38.
30. D. Wheatley, *The Deception Planners*, p.20.
31. R. Lewin, *The Chief,* p.52.

32. M. Howard, *British Intelligence in the Second World War: vol. 5, Strategic Deception*, p.33.

33. C. Cruickshank, *Deception in World War II*, pp. 19–21.

34. D. Mure, pp.95–6.

35. Ibid., pp.82–3, 93–4.

36. D. Mure, *Practise to Deceive*, p.40.

37. CAB 154/1 'A' Force War Diary, vol. 1, p.43.

38. PRO WO 201/2024.

39. CAB 154/1, pp.35, 51–2, 87.

40. CAB 154/2 'A' Force War Diary, vol. 2, p.19.

41. D. Mure, *Master of Deception*, p.81.

42. M. Howard, p.42.

43. D. Mure, *Master of Deception*, pp.120, 130.

44. B. Horrocks, *A Full Life*, p.125.

45. PRO WO 201/434.

46. B. H. Liddell Hart, *The Rommel Papers*, p.304.

CHAPTER 6. STRATEGIC DECEPTION

1. R. Patai, *The Arab Mind*, p.311.

2. J. Amos, 'Deception and the 1973 Middle East War', in D. C. Daniel and K. L. Herbig (eds.), *Strategic Military Deception*, pp.317–18.

3. W. Byford-Jones, *The Lightning War*, pp.33–4.

4. J. Amos, p.319.

5. *Sunday Times Insight Team, The Yom Kippur War*, pp.399–404.

6. J. Amos, p.320–1.

7. G. Meir, *My Life*, p.409.

8. See H. Trefousse, *Pearl Harbor: The Continuing Controversy*, chapter 2. A recent exposition of this theory can be found in R. B. Stinnett, *Day of Deceit: The Truth about FDR and Pearl Harbor*, New York, Free Press, 1999.

9. R. K. Betts, *Surprise Attack*, p.134.

10. *The Observer* (9 December 1979).

11. J. Hughes-Wilson, *Military Intelligence Blunders*, pp.68–9.

12. R. Wohlstetter, *Pearl Harbor: Warning and Decision*, pp.379–80.

13. B. Whaley, *Codeword Barbarossa*, p.242.

14. *Pravda* (22 June 1989), p.3.

15. J. Barros and R. Gregor, *Double Deception: Stalin, Hitler and the Invasion of Russia,* pp.16–7. This is an excellent and comprehensive study of the subject.

16. H. E. Salisbury, *The 900 Days: The Siege of Leningrad,* pp.57, 70.

17. R. H. S. Stolfi, 'Barbarossa: German Grand Deception and the Achievement of Strategic and Tactical Surprise against the Soviet Union, 1940–41', in D. C. Daniel and K. L. Herbig (eds.), *Strategic Military Deception,* p.201.

18. Ibid., p.195.

19. P. Fleming, *Operation Sealion,* p.296.

20. J. Barros and R. Gregor, pp.19–21.

21. R. H. S. Stolfi, p.196.

22. H. Trevor-Roper, *Hitler's War Directives*, p.57.

23. R. H. S. Stolfi, p.197.

24. NA H. Weberstedt, 'Public Deception Regarding the Russian Campaign', MS. P-044c, p.10; H. von Greiffenberg, 'Deception and Cover Plans', MS. P-044a.

25. R. H. Stolfi, pp.199–200.

26. G. Hagglof, *Samtida Vittne, 1940–45*, pp.96–7.

27. J. Barros and R. Gregor, pp.52–3, 56–8.

28. Ibid., p.59.

29. Ibid., pp.25–7.

30. R. H. S. Stolfi, p.201–2.

31. C. Cruickshank, *Deception in World War II*, p.209.

32. Sir W. S. Churchill, *The Second World War,* vol. 3, *The Grand Alliance,* p.357.

33. I. Akhmedov, *In and Out of Stalin's GRU; A Tartar's Escape from Red Army Intelligence, pp.136–7.*

34. J. Barros and R. Gregor, pp.122–7.

35. Ibid., pp.150–51.

36. J. Hughes-Wilson, pp.53–4.

37. G. Hilger and A. G. Meyer, *The Incompatible Allies: A Memoir–History of German Soviet Relations, 1918–1941,* pp.328–31.

38. H. R Trevor-Roper (ed.), *Hitler's Secret Conversations, 1941–1944, p.397.*

39. J. Barros and R. Gregor, p.164.

40. D. McLachlan, *Room 39: Naval Intelligence in Action, 1939–45, p.242.*

41. National Archives, Washington DC, E. F. J. Raus, 'Strategic Deception', MS. P-044b, p.5.

42. J. Erickson, *Hitler's War with Stalin: The Road to Stalingrad,* pp.88–9.

43. R. H. S. Stolfi, pp. 207–8.

44. K. Macksey, *Keselring: The Making of the Luftwaffe,* p.82.

45. H. Plocher, *The German Air Force versus Russia, 1941,* pp.33–5.

46. Bundesarchiv, Freiburg, Gen. Kdo LVIII Pz Korps, Kriegstagebuch 1, 19.02.1933–1.10.1941, p.19, 1568/1.
47. R. H. S. Stolfi, p.210–12.
48. E. E. J. Raus, *Strategic Deception*, p.5.
49. J. Barros and R. Gregor, pp.178–9.
50. Ibid., pp.199–200.
51. Ibid., pp.204–7.
52. D. Volkogonov, *Stalin: Triumph and Tragedy*, pp.407–9.
53. M. Howard, *British Intelligence in the Second World War, vol. 5: Strategic Deception*, p.22.
54. N. Wild, foreword to D. Mure, *Master of Deception*, p.11.
55. M. Howard, pp.3–10.
56. J. C. Masterman, *The Double Cross System*, p.9.
57. PRO CAB 121/105, SIC file A/ Policy/Deception/1, COS (41) 344th meeting, 7 October; JAPANESE (41) 819, 8 October.
58. PRO AIR 20 3693, 21.04.1942.
59. PRO CAB 154/2 'A' Force War Diary, vol. 2, p.60.
60. PRO CAB 121 /105 Wavell telegram to Prime Minister 12461 /C of 21 May 1942.
61. D. Wheatley, *The Deception Planners*, p.229.
62. Ibid., p.64.
63. M. Young and R. Stamp, *Trojan Horses*, p.16.
64. D. Mure, Master of Deception, pp.92–3.
65. Ibid., pp.72–5.
66. PRO CAB 154/100 Historical Record of Deception in the War against Germany and Italy (Sir Ronald Wingate's Narrative), vol. 1, p.111.

67. D. Mure, *Master of Deception,* pp.108–13.

68. M. Howard, pp.57–61.

69. L. F. Ellis, *Victory in the West,* vol. 1., p.10.

70. M. Howard, p.75.

71. H. Greiner and P. Schramm, *Kriegstagebuch des Oberkommando der Wehrmacht,* vol. 3, p.1219.

72. Pujol wrote an entertaining account of his life as a double agent: *Garbo: The Personal Story of the Most Successful Double Agent Ever,* London, Weidenfeld and Nicolson, 1985. Certainly he had no doubt of his own importance.

73. H. Greiner and P. Schramm, vol. 2, p.770.

74. M. Howard, pp.80–1.

75. H. Greiner and P. Schramm, vol. 3, p.1037.

76. D. Mure, *Master of Deception,* p.199.

77. NA RG 331 SHAEF/203/DX/Int.

78. PRO WO 199 464 79A.

79. H. Greiner and P. Schramm, vol. 3, p.1024.

80. Letter to Noël Wild (24 September 1945), quoted in D. Mure, *Master of Deception,* pp.234–6.

81. M. Howard, p.84.

82. J. B. Dwyer, *Seaborne Deception,* p.26.

83. PRO CAB 154/3 'A' Force War Diary, vol. 3, pp.69–70.

84. M. Howard, p.92.

85. PRO CAB 80 78, no. 16.

86. D. Mure, *Practise to Deceive,* pp. 102–3.

87. Ibid., pp.46, 53–5.

88. IWM MI 14/522/2, Kurze Feind Beurteilung West (KFW) 982 of 25.07.1943.

89. D. Mure, *Master of Deception,* p.210.

Chapter 7. Naval Deception

1. Vegetius, *Epitome of Military Science*, IV, 37, pp.135–6.
2. A. Wiel, *The Navy of Venice*, pp.172–4.
3. See D. Thomas, *Cochrane: Britannia's Last Sea-King, and I. Grimble, The Sea Wolf: The Life of Admiral Cochrane*.
4. S. Foote, *The Civil War: A Narrative*, vol. 2, p.125.
5. Ibid., vol. 3, pp.380–6. For a full account of the career of the *Alabama*, see C. M. Robinson III, *Shark of the Confederacy: The Story of the CSS Alabama*, London, Leo Cooper, 1995.
6. Sir J. Corbett, *Naval Operations*, vol. 1, p.177.
7. J. Terraine, *White Heat: The New Warfare, 1914–18*, p.253.
8. See Sir R. Bacon, *The Concise Story of the Dover Patrol*, London, Hutchinson, 1932.
9. G. Hartcup, Camouflage: *A History of Concealment and Deception in War*, pp.40–8.
10. For a detailed and very well-illustrated description of camouflage during the Second World War, including naval camouflage, see R. M. Stanley, *To Fool a Glass Eye*, p.53.
11. S. Reit, *Masquerade: The Amazing Camouflage Deceptions of World War II*, pp. 195–7.
12. R. M. Stanley, chapter 3.
13. J. Terraine, p.262. See also: E. Keble Chatterton, *Q-Ships and Their Story*, London, Sidgwick and Jackson, 1923. Written as it was so soon after the war, it mentions the shooting of the crew of U-27 in the most euphemistic terms, but is interesting nevertheless.

14. D. Woodward, *The Secret Raiders*, pp.138–44. This books gives a complete account of the operations of German disguised raiders. See also Rogge's own account, *The German Raider Atlantis*.

15. D. Pope, *The Battle of the River Plate*, pp.4–5.

16. D. Owen, *Battle of Wits: A History of Psychology and Deception in Modern Warfare*, p.42.

17. D. Pope, pp.147–52.

18. D. Macintyre, *Narvik*, p.30.

19. Sir W. S. Churchill, *The Second World War*, vol. 1, *The Gathering Storm*, p.579.

20. D. Owen, p.43.

21. 'Nazi Conspiracy and Aggression' (part of the Nuremburg documents) VI, pp.914–15 (N.D. C-115).

22. D. Macintyre, p.47.

23. W. Shirer, *The Rise and Fall of the Third Reich*, pp.698–9.

24. M. R. D. Foot, *Resistance*, p.241.

25. D. Owen, pp.43–5.

26. D. Van de Vat, *The Atlantic Campaign*, pp.250–6.

27. A. Weale, *Secret Warfare*, pp.1–3. For a full account of this daring action, see J. G. Dorrian, *Storming St Nazaire*, London, Pen & Sword, 1998.

28. S. Foote, vol. 3, pp.716–19.

27. M. Howard, *British Intelligence in the Second World War, pp.223–30.*

28. J. C. Masterman, *The Double Cross System*, pp.182–4.

29. For a full account of the US Navy's deception operations in the Pacific theatre, see K. L. Herbig, 'American

Strategic Deception in the Pacific' in M. I. Handel (ed.), *Strategic and Operational Deception in the Second World War*.

30. W. J. Homes, *Double-Edged Secrets*, pp.85–95; T. Tuleja, *Climax at Midway*, pp.51–2.

31. M. Fuchida and M. Okumiya, *Midway: The Battle that Doomed Japan*, pp.129–30.

32. J. B. Dwyer, Seaborne Deception: *The History of the US Navy Beach Jumpers*, pp.3–5.

33. M. Young and R. Stamp, *Trojan Horses: Deception Operations in the Second World War*, p.168.

34. J. B. Dwyer, p.20.

35. NA RG 218, 'Cover and Deception Plans' CSS 434/2, 06.01.1944.

36. J. B. Dwyer, p.73.

37. J. B. Dwyer, pp.127–30.

Chapter 8. Deception in Air Operations

1. M. Young and R. Stamp, *Trojan Horses: Deception Operations in the Second World War*, p.112.

2. R. M. Stanley, *To Fool a Glass Eye*, pp.73–8.

3. D. Mure, *Master of Deception*, p.98.

4. R. M. Stanley, pp.78–80.

5. J. Munson (ed.), *Echoes of the Great War: The Diary of Reverend Andrew Clark, 1914–19, p.153*.

6. PRO AIR 2/2878.

7. PRO AIR 2/6021.

8. S. Reit, *Masquerade: The Amazing Camouflage Deceptions of World War II*, p.52.

9. PRO AIR 2/3212.

10. PRO AIR 8/317.

11. PRO AIR 41/3 AHB Decoys and Deceptions.

12. PRO AIR 41/46 Air Historical Branch History of No. 80 (Signals) Wing.

13. S. Reit, pp.53–6.

14. H. R. Trevor-Roper, *Hitler's War Directives, 1939–1945*, p.79.

15. S. Reit, p.59.

16. Ibid., p.61.

17. M. E. DeLonge, *Modern Airfield Planning and Concealment*, New York, Pitman, 1943, p.135.

18. L. Brettingham, *Royal Air Force Beam Benders; No. 80 (Signals) Wing, 1940–45*, p.105.

19. PRO AIR 2/4759.

20. PRO AIR 26/580.

21. PRO AIR 26/583.

22. G. Pawle, *The Secret War*, pp.187–90.

23. S. Reit, p.208.

24. H. K. Smith, *Last Train from Berlin*, p.137.

25. R. M. Stanley, p.176.

26. H. W. Flannery, *Assignment to Berlin*, pp.293–4.

27. R. M. Stanley, p.25.

28. A full account is given in R. V. Jones, *Most Secret War*, chapter 11.

29. L. Brettingham, pp.17–18.

30. PRO AIR 41/46 pp.16–17.

31. Ibid., p.22.

32. W. Murray, *The Luftwaffe, 1933–45: Strategy for Defeat*, p.80.

33. D. Owen, *Battle of Wits: A History of Psychology and Deception in Modern Warfare*, p.62.
34. M. W. Bowman and T. Cushing, *Confounding the Reich*, p.63.
35. M. Dewar, *The Art of Deception in Warfare*, pp.151–3.
36. D. Owen, pp.64–5.
37. G. Hartcup, *Camouflage: A History of Concealment and Deception in War*, p.74.
38. M. Dewar, p.159.
39. W. Murray, p.214.
40. M. W. Bowman and T. Cushing, p.82.
41. D. Owen, pp.66–8.
42. M. W. Bowman and T. Cushing, pp.150–3.
43. R. M. Stanley, p.71.
44. E. H. Sims, *American Aces*, p.184.
45. W. Murray, p.300.
46. M. Howard, *British Intelligence in the Second World War: vol. 5, Strategic Deception*, p.170.
47. R.V. Jones, p.421.
48. M. Howard, pp.182–3.
49. D. Richardson, *Stealth Warplanes*, p.138.

CHAPTER 9. OPERATION BODYGUARD

1. N. Lewis, *Channel Firing*, p.99.
2. General order to US Seventh Army before the Sicily landings, 1943.
3. A. Cave Brown, *Bodyguard of Lies*, p.426.
4. M. van Creveld, *Supplying War*, p.208.
5. A. Cave Brown, p.426.

6. C. Wilmot, *The Struggle for Europe*, p.189.

7. B. H. Liddell Hart, *The Other Side of the Hill*, p.395.

8. C. Cruickshank, *Deception in World War II*, p.85.

9. PRO WO 219/308 (07.10.1943).

10. PRO CAB 80/76.

11. PRO CAB 122/1251.

12. SHAEF Ops 'B' document #18209, quoted in A. Cave Brown, p.460.

13. R. Kershaw, *D-Day: Piercing the Atlantic Wall*, p.20.

14. M. Young and R. Stamp, *Trojan Horses: Deception Operations in the Second World War*, pp.154–5.

15. PRO WO 199/1378, 26.03.1944.

16. R. Hesketh, *Fortitude*, p.37.

17. PRO WO 219/2221.

18. A. Cave Brown, p.465.

19. PRO WO 219/2220.

20. D. Owen, *Battle of Wits*, p.119.

21. C. Cruickshank, p.135.

22. PRO FO 188/446.

23. IWM MI 14/499, KFW 1279, 28.05.1944.

24. IWM Al 1828/1.

25. IWM KFW 1276, 25.05.1944.

26. H. Greiner and P. Schramm, *Kriegstagebuch des Oberkommando der Wehrmacht*, vol. 4, p.298.

27. D. Owen, p.120.

28. IWM MI 14/522/-2, KFW 978, 21.07.1943.

29. PRO CAB 146/57.

30. H. Greiner and P. Schramm, vol. 4, p.81.

31. R. Hesketh, pp.61–2, 157–9.

32. IWM AL 1828/2, 05.05.1944.

33. C. Cruickshank, p.157.
34. M. Howard, p.152.
35. PRO CAB 154/4, p.96.
36. M. Howard, p.125.
37. H. R. Trevor-Roper, *Hitler's War Directives, 1939–1945*, p.149.
38. B.H. Liddell Hart, pp.398–401.
39. C. Wilmot, p.191.
40. Earl Mountbatten of Burma, quoted in T. Robinson, *Dieppe: The Shame and the Glory*, p.15.
41. W. Murray, *The Luftwaffe, 1933–45: Strategy for Defeat*, p.135.
42. G. Perrault, *The Secret of D-Day*, p.165; H. Guderian, *Panzer Leader*, p.331.
43. R. Kershaw, p.35.
44. M. Howard, p.119.
45. C. Wilmot, p.200.
46. Ibid., p.207.
47. G. Perrault, p.220.
48. C. Cruickshank, p.171.
49. D. Owen, pp.124–7; D. Mure, *Master of Deception*, p.255.
50. R. Hesketh, p.25.
51. S. Reit, *Masquerade: The Amazing Camouflage Deceptions of World War II*, p.38.
52. R. Hesketh, pp.36–7.
53. PRO AIR 37/882.
54. A. Cave Brown, p.604.
55. R. Hesketh, pp.83–4.

56. S. Reit, pp.35–40.
57. M. Young and R. Stamp, pp.137–8.
58. M. Howard, pp.123–5.
59. J. C. Masterman, *The Double Cross System, 1939–45*, p.156.
60. D. Owen, pp.129–30.
61. J. Haswell, *The Intelligence and Deception of the D-Day Landings*, pp.150–1.
62. B. H. Liddell Hart, p.396.
63. D. Owen, p.134.
64. W. Warlimont, *Inside Hitler's Headquarters*, p.422.
65. D. Owen, pp.139–5.
66. PRO AIR 14/725.
67. M. Young and R. Stamp, pp.78–80.
68. M. Howard, p.188.
69. O. N. Bradley, *A Soldier's Story*, p.344.
70. G. Perrault, p.239.

Chapter 10. Maskirovka

1. Y. A. Yefrimov and S. G. Chermashentsev, 'Maskirovka', in *Soviet Military Encyclopedia*, vol. 5, Moscow, Voyenizdat, 1978, pp. 175–7.
2. M. Dewar, *The Art of Deception in Warfare*, p.83.
3. F. Myshak, 'Modern Camouflage', *Teckhnika-Molodeghi* (March 1968), pp.1–3.
4. P. Mel'nikov, 'Operational Maskirovka', in *Voyenno-Istoricheskiy Zhurnal* (April 1982), p.18.
5. G. K. Zhukov, *The Memoirs of Marshal Zhukov*, p.156.

6. E. Ziemke, 'Stalingrad and Belorussia: Soviet Deception in World War II', in D. C. Daniel and K. L. Herbig (eds.), *Strategic Military Deception*, p.243.

7. D. Glantz, *Soviet Military Deception in the Second World War*, pp.21–3.

8. F. Halder, *Kriegstagebuch*, vol. 7, Kolhammer, Stuttgart, 1962, pp.168–9.

9. Ibid., p.188.

10. A. Seaton, *The Battle for Moscow*, p.198.

11. E. Ziemke, 'Stalingrad and Belorussia', pp.246–8.

12. D. Glantz, pp.107–8.

13. Ibid., pp.99–103.

14. E. Ziemke, 'Stalingrad and Belorussia', p.250.

15. E. Ziemke, *Stalingrad to Berlin: The German Defeat in the East*, p.46.

16. E. Ziemke, 'Stalingrad and Belorussia', p.248.

17. L. B. Ely, *The Red Army Today*, p.87.

18. K. K. Rokossovskiy, *A Soldier's Duty*, p.152.

19. D. Glantz, p.113.

20. E. Ziemke, 'Stalingrad and Belorussia', p.249.

21. Ibid., pp.252–3.

22. A. Clark, *Barbarossa: The Russian-German Conflict, 1941–45*, p.272.

23. W. Görlitz, *Paulus and Stalingrad*, p.197.

24. Ibid., p.229.

25. NA 'AOK Ia Kriegstagebuch Nr. 14', 19 November 1942, in Sixth Army, File No. 33224/2.

26. D. Glantz, pp.118–19.

27. Ibid., p.561.

28. Ibid., pp.244–7.

29. Ibid., pp.267–8.
30. *Field Regulations of the Red Army 1944,* translation by the Office of the Assistant Chief of Staff G-2, GSUSA, published by JPRS, 1985.
31. G. K. Zhukov, p.525.
32. E. Ziemke, 'Stalingrad and Belorussia', p.260.
33. S. Shtemenko, *The Soviet General Staff at War, 1941–45,* pp.300–1.
34. K. K. Rokossovskiy, p.237.
35. E. Ziemke, 'Stalingrad and Belorussia', p.261.
36. NA 'OKH, GenStdH, FHO Nr. 1794/44, 2–3.6.44', in Foreign Armies East, File No. H 3/185.
37. E. Ziemke, 'Stalingrad and Belorussia', p.267.
38. E. Ziemke, *Stalingrad to Berlin,* pp.315–16.
39. S. Shtemenko, p.234.
40. E. Ziemke, 'Stalingrad and Belorussia', pp.268–9.
41. J. J. Dziak, 'Soviet Deception: The Organizational and Operational Tradition', in B. D. Dailey and P. J. Parker (eds.), *Soviet Strategic Deception,* pp.10–11. See also D. Sevin, 'Operation Scherhorn', *Military Review* (March 1966), pp.35–43.
42. D. Glantz, p.475–88.
43. Ibid., p.498.
44. M. Dewar, pp.88–9.
45. P. H. Vigor, *Soviet Blitzkrieg Theory*, p.105.
46. D. Glantz, pp.544–6.
47. P. H. Vigor, pp.108–10.
48. D. Glantz, pp.564, 569–70.
49. Ibid., pp.559–60.
50. P. H. Vigor, pp.149–50.

51. C. L. Smith, *Soviet Maskirovko*, www.airpower. maxwell.af.mil/airchronicles/apj/apj88/smith.html, pp.10–11.

52. L. Freedman and E. Karsh, *The Gulf Conflict*, p.367.

53. C. L. Smith, pp.5–8.

54. C. N. Donnelly, 'The Human Factor in Soviet Military Policy', *Military Review* (March 1985), p.21.

55. N. Trulock III, 'The Role of Deception in Soviet Military Planning', in B. D. Dailey and P. J. Parker (eds.), *Soviet Strategic Deception*, p.282.

56. R. Beaumont, *Maskirovka: Soviet Concealment, Camouflage and Deception*, p.31.

57. Ibid., p.34.

58. A. Postovalov, 'Modelling the Combat Operations of the Ground Forces', *Voyennaya Mysl,* no. 3 (1969).

59. P. H. Vigor, 'Doubts and Difficulties Confronting a Would-Be Soviet Attacker', *RUSI Journal* (June 1980), pp.32–8.

60. J. Valenta, 'Soviet Views of Deception', in D. C. Daniel and K. L. Herbig (eds.), *Strategic Military Deception*, pp.339–40.

61. See J. Valenta, *Soviet Intervention in Czechoslovakia, 1968: Anatomy of a Decision*.

62. G. F. Krivosheev (ed.), *Removing the Secret Seal: Casualty Figures of the Armed Forces in War, Combat Action and Military Conflicts, Moscow, Voyenizdat, 1993*, pp.397–8.

63. L. Grau (ed.), *The Bear Went over the Mountain: Soviet Combat Tactics in Afghanistan*, pp.199–200.

64. P. H. Vigor, *Soviet Blitzkrieg Theory*, p.141.

65. J. Valenta in D. C. Daniel and K. L. Herbig, pp.345–7.

66. M. Dewar, p.190.
67. L. Grau, pp.61–4.

Chapter 11. Deception in Counter-Revolutionary and Irregular Warfare

1. E. Montagu, *The Man Who Never Was*, pp.109–10.
2. M. R. D. Foot, *Resistance*, p. 181.
3. M. Young and R. Stamp, *Trojan Horses: Deception Operations in the Second World War, p.198.*
4. D. Brown, *The Fetterman Massacre.*
5. C. E. Callwell, *Small Wars*, p.54.
6. M. Dewar, *The Art of Deception in Warfare*, pp.181.
7. C. E. Callwell, p.55.
8. Ibid., p.174.
9. Ibid., pp.175–6.
10. Mao Tse-tung, On Guerrilla *Warfare*, p.41.
11. W. W. Whitson, *The Chinese High Command*, pp.173–5.
12. F. Kitson, *Low Intensity Operations*, p.29.
13. K. Jeffery, 'Colonial Warfare, 1900–39', in C. McInnes and G. D. Sheffield (eds.), *Warfare in the Twentieth Century: Theory and Practice*, p.32.
14. See C. Sykes, *Orde Wingate*, London, Collins, 1959, pp.141, 149–58.
15. M. Dewar, pp.182–3.
16. F. Kitson, p.95.
17. P. Melshen, 'Pseudo-Operations', pp.23–7, 34. This excellent study of the subject is unfortunately not available in published form.

18. R. Jackson, *The Malayan Emergency,* p.14.
19. PRO DEFE 11/36, 12.05.1950.
20. P. Melshen, pp.43–4.
21. PRO DEFE 11/42, 30.11.1950.
22. D. Owen, *Battle of Wits,* pp.158–64.
23. P. Melshen, pp.59–60.
24. R. B Edgerton, *Mau Mau: An African Crucible,* pp. 42–55. See also F. Furendi, *The Mau Mau War in Perspective,* London, James Curry, 1989.
25. F. Kitson, *Gangs and Counter-Gangs,* pp.73–4. This is a personal account of Kitson's experiences in Kenya. He later wrote *Low Intensity Operations* as a result of a defence scholarship in which pseudo-operations are placed within his framework for counter-insurgency (pp.95–102, 187–97). See also *Bunch of Fives,* a summation of his career.
26. Ibid., pp.82–4.
27. Ibid., pp.93–4, 102–4, 121–2, 186–7, 209.
28. P. Melshen, pp. 108–9.
29. R. Reid Daley, *Selous Scouts: Top Secret War,* Alberton, RSA, Galago Publishing, 1982; and P. Stiff, *Selous Scouts,* Alberton, RSA, Galago Publishing, 1984. Both give somewhat self-congratulatory and glorified accounts (Reid Daley was a former commanding officer of the unit).
30. The Q-patrols of Cyprus are described in G. Grivas, *Guerrilla Warfare and EOKA's Struggle* (trans. A. A. Pallis), London, Longman's, 1964, pp.50–2.
31. T. Geraghty, *Who Dares Wins,* pp.401–2.

32. M. Barthorp, *Crater to the Creggan: A History of the Royal Anglian Regiment, 1964–74*, p.39.

33. P. Melshen, p.149.

34. A. Home, *A Savage War of Peace*, p.390.

35. See B. J. Kerkvliet, *The Huk Rebellion: A Study of Peasant Revolt in the Philippines*, Berkeley, CA, University of California Press, 1977; and E. Lachica, *The Huk: Philippine Agrarian Society in Revolt*, New York, Praeger, 1971.

36. P. Melshen, p.220.

37. R. H. Spector, *Advice and Support: United States Army in Vietnam, The Early Years, 1941–60*, pp.221–3, 282–6, 296–302.

38. A. Krepinevich, *The Army and Vietnam*, pp.4–5.

39. *Counter-Insurgency Operations: A Handbook for the Suppression of Communist Guerrilla/Terrorist Operations*, Norfolk, VA, Armed Forces Staff College, 01.03.1962, p.34.

40. A personal account of these operations can be found in J. E. Acre, *Project Omega: Eye of the Beast*, Hellgate, 1999.

41. P. Melshen, pp.265–72.

42. W. C. Westmoreland, *A Soldier Reports*, pp.164–6. See also L. W. Walt, *Strange War, Strange Strategy*, New York, Funk and Wagnalls, 1970, for the Marine commander's view.

43. P. Melshen, p.298.

44. J. B. Dwyer, *Seaborne Deception: The History of US Navy Beach Jumpers*, pp.114–15.

45. J. R. Arnold, *Tet Offensive, 1968*, p.21.
46. R. Beaumont, *Maskirovka: Soviet Camouflage, Concealment and Deception*, p.42.
47. P. Melshen, p.301.
48. M. Dewar, p.190.
49. W. Colby, *Lost Victory*, p.264.
50. See G. Lewy, 'Deception and Revolutionary Warfare', in D. A. Charters and M. A. J. Tugwell (eds.), *Deception Operations: Studies in the East West Context.*
51. J. Arnold, p.12.
52. J. Hughes-Wilson, *Military Intelligence Blunders*, pp.175–80.
53. D. Oberdorfer, *Tet! The Turning Point in the Vietnam War*, p.119.
54. J. Hughes-Wilson, p.198.
55. Ibid., 210–15.
56. P. Bishop and E. Mallie, *The Provisional IRA*, pp.177–8.
57. P. Melshen, p. 183.
58. M. Urban, *Big Boys Rules: The Secret Struggle Against the IRA*, p.37–49.
59. P. Bishop and E. Mallie, pp.239–42.
60. J. Parker, *Death of a Hero*, pp.75–8.

Chapter 12. The Future of Deception

1. 'Journalists Must Always Fight Spin', *The Independent* (17 January 2000).
2. J. Haswell, *The Tangled Web: The Art of Tactical and Strategic Deception*, p.124.
3. Ibid., p.111–12.

4. L. Freedman and E. Karsh, *The Gulf Conflict*, p.386.

5. J. Blackwell, Thunder *in the Desert: The Strategy and Tactics of the Persian Gulf War*, pp.156–7.

6. L. Freedman and E. Karsh, p.387.

7. H. N. Schwarzkopf, *It Doesn't Take a Hero*, pp.381, 440.

8. L. Freedman and E. Karsh, p.395.

9. N. Freidman, *Desert Victory: The War for Kuwait*, p.221.

10. D. Steele, 'Tanks and Men: Desert Storm from the Hatches', *Army* (June 1991).

11. *Los Angeles Times* (8 August, 17 August 1990, 1 January 1991); *Washington Post* (8 August 1990); *New York Times* (25 January 1991).

12. 'Marines Are at Sea and Unhappy', *Washington Times* (18 January 1991); 'To the Shores of Kuwait', *Newsweek* (11 February 1991).

13. 'Stolen Computer Contained Gulf Deception Plan', *The Times* (13 March 1991).

14. A. M. Codevilla, 'Space, Intelligence and Deception', in B. D. Dailey and P. J. Parker (eds.), *Soviet Strategic Deception*, pp.471–3.

15. Ibid., pp.474–5.

16. N. Trulock III, 'The Role of Deception in Soviet Military Planning', in B. D. Dailey and P. J. Parker (eds.), *Soviet Strategic Deception*, p.289.

17. R. Beaumont, *Maskirovka: Soviet Camouflage, Concealment and Deception*, p.30.

18. C. L. Smith, *Soviet Maskirovko*, www.airpower.max-well.af.mil/airchronicles/apj/apj88/smith.html, p.10.

19. A. P. Wavell, *The Good Soldier*, pp.121–56.
20. D. C. Daniel and K. L. Herbig, 'Propositions on Military Deception', in *Strategic Military Deception*, p.14.
21. 'Army Goes to War with Platoons of Holograms', *Sunday Telegraph* (11 May 1997).
22. V. Shcedrov, 'Camouflaging Troops during Regrouping and Manoeuvre', *Voennaya Mysl* (June 1966).
23. 'US "Lost Count of Uranium Shells Fired in Kosovo" ', *The Independent* (22 November 1999).
24. 'Nato Hits Were Only a Tenth of Those Claimed, Says US Air Force', *The Independent* (8 May 2000).
25. R. Beaumont, p.18.
26. A. M. Codevilla, pp.468–9.
27. Ibid., p.10.
28. *Cornerstones of Information Warfare*, www.af.mil/lib/comer.html, p.2.
29. M. R. D. Foot, *Resistance*, p.28.
30. A. P. Wavell, p.47.
31. K. J. Müller, 'A German Perspective on Allied Deception Operations in the Second World War', in M. I. Handel (ed.), *Strategic and Operational Deception in the Second World War.*
32. D. Mure, *Master of Deception*, p.87.

BIBLIOGRAPHY

Akhmedov, I., *In and Out of Stalin's GRU: A Tartar's Escape from Red Army Intelligence,* Frederick, SD, University Publications of America, 1984

Arnold, J. R., *Tet Offensive 1968,* Oxford, Osprey, 1990

Babington-Smith, C., *Air Spy,* New York, Harper, 1957

Baden-Powell, R. S. S., *Lessons from the 'Varsity of Life,* London, Pearson, 1933

Barkas, G., *The Camouflage Story,* London, Cassell, 1952

Barker, A. J., *Fortune Favours the Brave: The Battle of the Hook,* Korea, 1953, London, Leo Cooper, 1974

Barros, R., *and Gregor, J., Double Deception: Stalin,* Hitler and the Invasion of Russia, DeKalb, Northern Illinois University Press, 1995

Barthorp, M., *Crater to the Creggan: A History of the Royal Anglian Regiment, 1964–74,* London, Leo Cooper, 1976

Beaumont, R., *Maskirovka: Soviet Camouflage, Concealment and Deception,* Stratech Studies, Texas A & M University, 1982

Beesley, P., *Room 40: British Naval Intelligence, 1914–18,* London, Hamish Hamilton, 1982

Betts, R. K., *Surprise Attack,* Washington DC, Brookings Institution, 1982

Bishop, P., and Mallie, E., *The Provisional IRA,* London, Transworld Corgi, 1988

Blackwell, J., *Thunder in the Desert: The Strategy and Tactics of the Persian Gulf War,* New York, Bantam Books, 1991

Bowman, M. W., and Cushing, T., *Confounding the Reich,* Sparkford, Patrick Stephens, 1996

Bradley, O. N., *A Soldier's Story,* New York, Henry Holt, 1951

Brettingham, L. M., *Royal Air Force Beam Benders: No. 80 (Signals) Wing, 1940–45,* Leicester, Midland, 1997

Brookes, A. J., *Photo Reconnaissance,* Shepperton, Ian Allen, 1975

Brown, D., *The Fetterman Massacre,* London, Barrie and Jenkins, 1972

Byford-Jones, W., *The Lightning War,* London, Robert Hale, 1967

Caesar, J., *The Conquest of Gaul* (trans. S. A. Handford), London, Penguin, 1982

Callwell, C. E., *Small Wars: A Tactical Textbook for Imperial Soldiers,* London, Greenhill, 1990

Cave Brown, A., *Bodyguard of Lies,* London, W. H. Allen, 1986

Chambers, J., *The Devil's Horsemen: The Mongol Invasion of Europe,* London, Cassell, 1988

Chandler, D. G., *Atlas of Military Strategy: The Art, Theory and Practice of War, 1618–1878,* Don Mills, Ontario,

Collier Macmillan Canada/Fortress Publications, 1980

—— *The Campaigns of Napoleon*, London, Weidenfeld and Nicolson, 1966

—— *Marlborough as Military Commander*, London, Batsford, 1973

Charters, D. A., and Tugwell, M. A. J., *Deception Operations: Studies in the East-West Context*, London, Brassey's, 1990

Churchill, Sir W. S., T*he Second World War* (6 vols.), London, Cassell, 1951

Clark, A., *Barbarossa: The Russian-German Conflict, 1941–45*, London, Hutchinson, 1965

Clausewitz, C. von, *On War*, Harmondsworth, Pelican, 1968

Colby, W. E., *Lost Victory*, Chicago, Contemporary Books, 1989

Corbett, Sir J., *Naval Operations*, London, Longmans, Green & Co., 1920

Cottrell, L., *Enemy of Rome*, London, Evans Bros., 1960

Cruickshank, C., *Deception in World War II*, Oxford, Oxford University Press, 1979

Dailey, B. D., and Parker, P. J. (eds.), *Soviet Strategic Deception*, Lexington, MA, D. C. Heath, 1987

Daniel, D. C., and Herbig, K. L. (eds.), *Strategic Military Deception*, Oxford, Pergamon, 1982

Davies, J. D. G., *Owen Glyn Dŵr*, London, Eric Partridge, 1934 De Arcangelis, M., *Electronic Warfare*, Poole, Blandford Press, 1985

Dear, I. C. B. (ed.), *The Oxford Companion to the Second World War,* Oxford University Press, 1995

Delaforce, T., *Monty's Marauders,* Brighton, Tom Donovan, 1997

—— *The Fighting Wessex Wyverns,* Stroud, Alan Sutton, 1994

Delbrück, H., *History of the Art of War within the Framework of Political History,* 4 vols., London, Greenwood Press, 1975–85

Dewar, M., *The Art of Deception in Warfare,* Newton Abbot, David & Charles, 1989

Dixon, N., *On the Psychology of Military Incompetence,* London, Jonathan Cape, 1976

Duffy, C., *Frederick the Great: A Military Life,* London, Routledge, 1985

Dupuy, R. E., and Dupuy, T. N., *The Collins Encyclopaedia of Military History,* 4th edn, London, Harper Collins, 1993

Dwyer, J. B., *Seaborne Deception: The History of US Navy Beach Jumpers,* New York, Praeger, 1992

Edgerton, R. B., *Mau Mau: An African Crucible,* London, I. B. Taurus, 1996

Ellis, L. E, *History of the Second World War: Victory in the West,* vol. 2, London, HMSO, 1968

Ely, L. B., *The Red Army Today,* Harrisburg, PA, Military Service Publishing Company, 1949

Erickson, J., *Hitler's War with Stalin: The Road to Stalingrad,* London, Weidenfeld and Nicholson, 1975

Flannery, H. W., *Assignment to Berlin,* London, Michael Joseph, 1942

Fleming, P., *Operation Sealion,* London, Pan, 1975

Foot, M. R. D., *Resistance,* London, Eyre Methuen, 1976

—— *MI9: Escape and Evasion, 1939–45,* London, Futura, 1979

Foote, S., *The Civil War: A Narrative,* 3 vols., London, Pimlico, 1992

Freedman, L., and Karsh, E., *The Gulf Conflict,* London, Faber & Faber, 1994

Freidman, N., *Desert Victory: The War for Kuwait,* Annapolis, VA, Naval Institute Press, 1991

Frontinus, S. J., *The Stratagems,* London, Loeb Classical Library, William Heinemann, 1925

Fuchida, M., and Okumiya, M., *Midway: The Battle that Doomed Japan,* Annapolis, VA, Naval Institute Press, 1955

Gardner, B., *Mafeking: A Victoria Legend,* London, Cassell, 1966

Geraghty, T., *Who Dares Wins: The Story of the SAS, 1950–1980,* London, Arms and Armour, 1980

Gilbert, A., *Sniper: One-on-One,* London, Sidgwick and Jackson, 1994

Glantz, D., *Soviet Military Deception in the Second World War,* London, Frank Cass, 1989

Görlitz, W., *Paulus and Stalingrad,* London, Methuen, 1963

Grau, L. W. (ed.), *The Bear Went over the Mountain: Soviet Combat Tactics in Afghanistan,* London, Frank Cass, 1998

Greiner, H., and Schramm, P. (eds.), *Kriegstagebuch des Oberkommando der Wehrmacht,* 4 vols., Frankfurt am Main, Bernard und Graefe, 1961–4

Grimble, I., *The Sea Wolf: The Life of Admiral Cochrane,* London, Blond & Briggs, 1978

Grinnell-Milne, D., *Mafeking,* London, The Bodley Head, 1957

Guderian, H., *Panzer Leader,* London, Futura, 1982

Hagglof, G., *Samtida Vittne, 1940–45,* Stockholm, Norstedt, 1972

Handel, M. I. (ed.), *Strategic and Operational Deception in the Second World War,* London, Frank Cass, 1987

—— (ed.), *Intelligence and Military Operations,* London, Frank Cass, 1990

Hartcup, G., *Camouflage: A History of Concealment and Deception in War,* Newton Abbot, David & Charles, 1979

Haswell, J., *The Intelligence and Deception of the D-Day Landings,* London, Batsford, 1979

—— *The Tangled Web: The Art of Tactical and Strategic Deception,* Wendover, John Goodchild, 1985

Henderson, N., *Prince Eugen of Savoy,* London, Weidenfeld and Nicolson, 1965

Hesketh, R., *Fortitude,* London, Little Brown, 1999

Hickey, M., *Gallipoli,* London, John Murray, 1995

Hilger, G., and Meyer, A. G., *The Incompatible Allies: A Memoir-History of German Soviet Relations 1918–1941,* New York, Macmillan, 1953

Holmes, R., *War Walks* 2, London, BBC, 1997

Homes, W. J., *Double-Edged Secrets,* Annapolis, VA, Naval Institute Press, 1979

Home, A., *The Price of Glory,* London, Penguin, 1993

—— *A Savage War of Peace,* London, Macmillan, 1996

Horrocks, Sir B., *A Full Life,* London, Collins, 1960

Howard, M., *British Intelligence in the Second World War,* vol. 5: *Strategic Deception,* London, HMSO, 1990

Hughes-Wilson, J., *Military Intelligence Blunders,* London, Robinson, 1999

Jackson, R., *The Malayan Emergency,* London, Routledge, 1991

James, E. C., *I Was Monty's Double,* London, Panther, 1958

Jones, R. V., *Most Secret War,* London, Hamish Hamilton, 1978

Katcher, P., *Great Gambles of the Civil War,* London, Arms and Armour, 1996

Kershaw, R. J., *D-Day: Piercing the Atlantic Wall,* Shepperton, Ian Allen, 1993

Kessler, L., *Kommando: Hitler's Special Forces in the Second World War,* London, Leo Cooper, 1995

Kitson, F., *Gangs and Counter-Gangs,* London, Barrie and Rockliff, 1960

—— *Low Intensity Operations,* London, Faber and Faber, 1971

—— *Bunch of Fives,* London, Faber and Faber, 1977

Krepinevich, A., *The Army and Vietnam,* Baltimore, MD, Johns Hopkins University Press, 1986

Lewin, R., *The Chief,* London, Hutchinson, 1980

—— *Ultra Goes to War,* London, Hutchinson, 1979

Lewis, N., *Channel Firing,* London, Penguin, 1990

Liddell Hart, B. H., *The Other Side of the Hill,* London, Cassell, 1948

—— *The Rommel Papers,* London, William Collins, 1953

Lucas, J., *Kommando: German Special Forces of World War Two,* London, Arms and Armour, 1985

—— *War on the Eastern Front, 1941–45: The German Soldier in Russia,* New York, Stein & Day, 1979

Luvaas, J. (ed.), *Frederick The Great on the Art of War,* New York, The Free Press, 1966

Machiavelli, N., *The Art of War,* Indianapolis, IN, Bobbs-Merrill Co. Inc., 1965

MacIntyre, D., *Narvik,* London, Evans Brothers, 1959

Macksey, K., *Kesselring: The Making of the Luftwaffe,* New York, David McKay, 1978

McInnes, C., and Sheffield, G. D. (eds.), *Warfare in the Twentieth Century: Theory and Practice,* London, Unwin Hyman, 1988

McKay, D., *Prince Eugene of Savoy,* London, Thames and Hudson, 1977

McLachlan, D., *Room 39: Naval Intelligence in Action, 1939–45,* London, Weidenfeld and Nicolson, 1968

Mao Tse-tung, 'Problems of Strategy in China's Revolutionary War', *Selected Works,* vol. 1, Peking, Foreign Languages Press, 1967

—— *On Guerrilla Warfare* (trans. A. Griffith), NY, Anchor Press, Doubleday, 1978

Maskelyne, J., *Magic: Top Secret,* London, Stanley Paul, 1949

Masterman, Sir J. C., *The Double Cross System in the War of 1939 to 1945,* London, Sphere, 1973

Meir, G., *My Life,* New York, Dell, 1975

Mellenthin, F. W. von, *Panzer Battles: A Study of the Employment of Armor in the Second World War* (trans.

H. Betzler), Norman, OK, University of Oklahoma Press, 1956

Melshen, P., 'Pseudo-Operations: The Use by British and American Armed Forces of Deception in Counter-Insurgencies, 1945–1973', PhD thesis, University of Cambridge, 1995

Meyer, H., *Kriegsgeschichte der 12. SS Panzer Division Hitlerjügend,* Osnabrück, Munin, 1982

Montagu, E., *The Man Who Never Was,* London, Evans Brothers, 1953

Moorehead, A., *Gallipoli,* London, Hamish Hamilton, 1956

Morgan, D., *The Mongols,* Oxford, Blackwell, 1986

Morris, D., *The Washing of the Spears,* London, Jonathan Cape, 1965

Morrocco, J., et al., *The Vietnam Experience: Thunder in the Air, 1941–68,* Boston, MA, Boston Publishing Co., 1984

Munson, J. (ed.), *Echoes of the Great War: The Diary of Reverend Andrew Clark,* 1914–19, Oxford, OUP, 1985

Mure, D., *Practise to Deceive,* London, William Kimber, 1977

—— *Master of Deception,* London, William Kimber, 1980

Murray, W., *The Luftwaffe, 1933–45: Strategy for Defeat,* Washington, Brassey's, 1996

Newark, T., Newark, Q., and Borsarello, J. F., *Brassey's Book of Camouflage,* London, Brassey's, 1996

Newton, S. H., *German Battle Tactics on the Russian Front, 1941–1945,* Atglen, PA, Schiffer, 1994

Oberdorfer, D., *Tet! The Turning Point of the Vietnam War,* New York, Da Capo, 1984

Oman, Sir C., *A History of the Art of War in the Middle Ages*, vol. 1, London, Methuen, 1978

Owen, D., *Battle of Wits: A History of Psychology and Deception in Modern Warfare*, London, Leo Cooper, 1978

Pakenham, T., *The Boer War*, London, Weidenfeld and Nicolson, 1979

Parker, J., *Death of a Hero*, London, Metro, 1999

Patai, R., *The Arab Mind*, New York, Scribner's Sons, 1973

Pawle, G., *The Secret War*, New York, William Sloane, 1957

Peddie, J., *Hannibal's War*, Stroud, Alan Sutton, 1997

Perrault, G., *The Secret of D-Day*, London, Arthur Barker, 1965

Pimlott, J., *Vietnam: The Decisive Battles*, London, Michael Joseph, 1990

Pitt, B., *The Crucible of War*, London, Jonathan Cape, 1980

Plocher, H., *The German Air Force versus Russia 1941*, New York, Amo Press, 1965

Polybius, *The Rise of the Roman Empire*, London, Penguin, 1979

Pope, D., T*he Battle of the River Plate*, London, Chatham, 1999

Reit, S., *Masquerade: The Amazing Camouflage Deceptions of World War II*, London, Robert Hale, 1979

Richardson, D., *Stealth Warplanes*, London, Salamander, 1989

Robinson, T., *Dieppe: The Shame and the Glory*, London, Hutchinson, 1962

Roetter, C., *Psychological Warfare*, London, Batsford, 1974

Rogge, B., and Frank, W., *The German Raider Atlantis*, London, Bantam, 1979

Rokossovskiy, K. K., *A Soldier's Duty*, Moscow, Progress, 1970

Roskill, S. N., *History of the Second World War: The War at Sea*, vol. 1, London, HMSO, 1956

Salisbury, H. E., *The 900 Days: The Siege of Leningrad*, London, Secker and Warburg, 1969

Saunders, H. St G., *The Red Beret: The Story of the Parachute Regiment at War, 1940–45*, London, Michael Joseph, 1950

Schick, I. T., (ed.), *The Uniforms of the World's Great Armies, 1700 to the Present*, London, Weidenfeld and Nicolson, 1978

Schwarzkopf, H. N., *It Doesn't Take a Hero*, London, Bantam, 1992

Seaton, A., *The Battle for Moscow*, New York, Playboy Press, 1971

Sellwood, A. V., *The Saturday Night Soldiers*, London, Wolfe Publishing, 1966 Shirer, W. L., *The Rise and Fall of the Third Reich*, London, Mandarin, 1991

Shtemenko, S. M., *The Soviet General Staff at War, 1941–45*, Moscow, Progress, 1986

Sims, E. H., *American Aces*, New York, Ballantine, 1958 Smith, H. K., *Last Train from Berlin*, London, Cresset, 1942

Spector, R. H., *Advice and Support: United States Army in Vietnam, The Early Years, 1941–60*, New York, The Free Press, 1985

Stanley, R. M., *To Fool a Glass Eye,* Shrewsbury, Airlife, 1998

Strachan, H., *European Armies and the Conduct of War,* London, Allen and Unwin, 1983

Stuart-Jones, E. H., *The Last Invasion of Britain,* Cardiff, University of Wales Press, 1950

Sunday Times Insight Team, *The Yom Kippur War,* London, Andre Deutsch, 1975

Sun Tzu, *The Art of War* (trans. S. Griffiths), Oxford, Oxford University Press, 1963

Taber, R., *The War of the Flea,* St Albans, Paladin, 1970

Taylor, G., *Infantry Colonel,* Upton upon Severn, Self Publishing Association, 1990

Terraine, J., *White Heat: The New Warfare, 1914–18,* London, Sidgwick and Jackson, 1982

Thomas, D., *Cochrane: Britannia's Last Sea-King,* London, Andre Deutsch, 1978

Thorau, P., *The Lion of Egypt: Sultan Baybars I and the Near East in the Thirteenth Century,* London, Longman, 1991

Trefousse, H. L., *Pearl Harbor: The Continuing Controversy,* Malabar, FL, Robert E. Krieger, 1982

Trevor-Roper, H. R. (ed.), *Hitler's Secret Conversations, 1941–1944,* New York, Farrar, Strauss and Young, 1953

—— *Hitler's War Directives,* London, Pan, 1966

Tuleja, T., *Climax at Midway,* New York, W. W. Norton, 1960

Tuchman, B., *The Guns of August: August 1914,* London, Constable, 1962

Urban, M., *Big Boys Rules: The Secret Struggle Against the IRA,* London, Faber and Faber, 1992

Valenta, J., *Soviet Intervention in Czechoslovakia, 1968: Anatomy of a Decision*, Baltimore, MD, Johns Hopkins University Press, 1979

Van Creveld, M., *Supplying War,* Cambridge, Cambridge University Press, 1977

Van der Vat, D., *The Atlantic Campaign,* London, Hodder and Stoughton, 1988

Vegetius, *Epitome of Military Science* (trans. N. P. Milner), Liverpool University Press, 1993

Vigor, P. H., *Soviet Blitzkrieg Theory,* London, Macmillan, 1983

Volkogonov, D., *Stalin: Triumph and Tragedy,* London, Weidenfeld and Nicolson, 1991

Ward, G. C., *with Bums, R., and Bums, K.,* The Civil War, *London, Bodley Head, 1991*

Warlimont, W., *Inside Hitler's Headquarters,* London, Weidenfeld and Nicolson, 1964

Watson, B. W., Bruce, G., Tsouras, P., Cyr, B. L., et al., *Military Lessons of the Gulf War*, London, BCA, 1991

Wavell, A. P., *The Palestine Campaigns,* London, Constable, 1928

—— *The Good Soldier,* London, Macmillan, 1948

Weale, A., *Secret Warfare,* London, Hodder and Stoughton, 1997

Westmoreland, W. D., *A Soldier Reports,* Garden City, NY, Doubleday, 1976

Whaley, B., *Codeword Barbarossa,* Cambridge, MA, MIT Press, 1973

—— *Stratagem: Deception and Surprise in War,* Cambridge, MA, MIT Press, 1969

Wheatley, D., *The Deception Planners,* London, Hutchinson, 1980

Whitson, W. W., *The Chinese High Command,* New York, Praeger, 1973

Wiel, A., *The Navy of Venice,* London, John Murray, 1910

Williams, G. A., *Owain Glyndŵr,* Cardiff, University of Wales Press, 1993

Wilmot, C., *The Struggle for Europe,* London, Collins, 1952

Wilson, F., *Regiments at a Glance,* London, Blackie and Sons, 1954

Wohlstetter, R., *Pearl Harbor: Warning and Decision,* Stanford, CA, Stanford University Press, 1962

Woodward, C. V. (ed.), *Mary Chesnut's Civil War,* New Haven, CT, Yale University Press, 1981

Woodward, D., *The Secret Raiders, London,* New English Library, 1975

Wyeth, J. A., *Life of General Nathan Bedford Forrest,* Edison, NJ, Blue and Grey Press, 1996

Young, M., and Stamp, R., *Trojan Horses: Deception Operations in the Second World War,* London, The Bodley Head, 1989

Zhukov, G. K., *The Memoirs of Marshal Zhukov,* London, Jonathan Cape, 197

Ziemke, E., *Stalingrad to Berlin: The German Defeat in the East,* Washington DC, US Government Printing Office, 1967